Nicholas Morgan

The Skull and Brain

Their indications of character and anatomical relations

Nicholas Morgan

The Skull and Brain
Their indications of character and anatomical relations

ISBN/EAN: 9783743345058

Manufactured in Europe, USA, Canada, Australia, Japa

Cover: Foto ©berggeist007 / pixelio.de

Manufactured and distributed by brebook publishing software (www.brebook.com)

Nicholas Morgan

The Skull and Brain

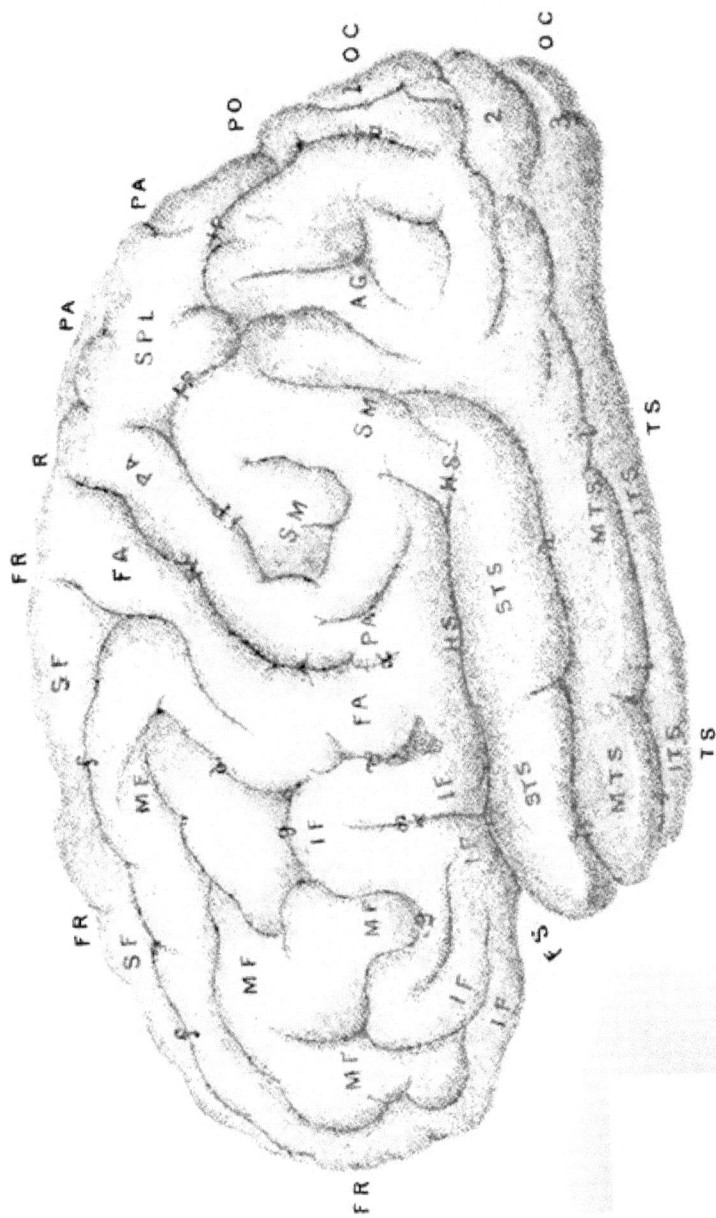

No 1

THE

SKULL AND BRAIN:

THEIR INDICATIONS OF

CHARACTER

AND

ANATOMICAL RELATIONS.

By NICHOLAS MORGAN,

Author of "Phrenology, and How to Use it in Analyzing Character," etc.

ILLUSTRATED BY

LITHOGRAPHIC AND WOOD ENGRAVINGS.

SPECIALLY GOT UP FOR THE WORK.

LONDON:

LONGMANS, GREEN & CO.

DEDICATION.

PREFACE.

SINCE the publication of " Phrenology, and How to Use it in Analyzing Character," the want of a more elementary work for the use of persons who have not time to read long treatises, and for beginners in the study of character, has often been pointed out to me ; and I began the present work with the aim of supplying such want. My first design was to answer some of the principal objections brought against phrenology by some high authorities in special departments of literature and science, so as to remove the chief stumbling-blocks out of the way, and then to do little more than give an epitome of the aforesaid work : but, after a few pages of the introduction were written, another plan was designed, the execution of which it is hoped, will be found more in keeping with the public requirements and general usefulness.

Excepting some paragraphs on the anatomy of the skull and brain, and of the temperaments, which are extracted from " Phrenology, and How to Use it," the work is entirely new, both in the matter and method of its treatment. The beaten track of phrenological literature has been left, and a new one struck out that appears more likely to aid phrenology in conciliating honourable opponents, and inducing them to give the subject that attention and impartial investigation which its supporters think it deserves, and its importance demands.

The custom is to speak of faculties, organs, and functions ; but, in the present case, excepting in the controversial parts of the work, in which I deemed it better to follow the usual course, I principally treat of signs of character, and how to translate their meanings without any reference to organs, etc.

Some of the more prominent objections to phrenology are reviewed,—especially those advanced by Mr. G. H. Lewes in his "History of Philosophy." Dr. Ferrier's researches in Cerebral Physiology receive a passing notice. The Will comes in for a considerable share of attention and Dr. Carpenter's exposition of it in "Mental Physiology" is examined. The topography of the cerebral convolutions, and Professor Turner's Cranio-Cerebral Map, showing the relations of the convolutions of the outer surface of the hemispheres to the skull, are pretty fully treated of, and illustrated by lithographic drawings, including one used by Professor Turner to illustrate the subject, which he has kindly permitted me to reproduce. The Professor divides each lateral half of the skull into ten areas, which should command the attention of phrenologists. Believing that these areas might be advantageously subdivided, I have increased them to seventeen, and given rules for determining their boundaries in the head, and their relative positions to the cerebral convolutions.

Size and quality as measures of power, and the
effects of temperament on the mental operations,
are concisely yet comprehensively treated ; likewise
how to examine heads and crania, so as to discern
their peculiarities, and to understand their indications. All other necessary instruction is also given
in the theory and practice of phrenology for the
student's guidance.

                                                N. M.

May 1875.

CONTENTS.

CHAPTER I.

OBJECTIONS TO PHRENOLOGY REVIEWED.

CHAPTER V.

THE RELATIONS OF THE OUTER CEREBRAL CONVOLUTIONS TO THE SKULL.

CHAPTER VI.

PLATE I.

THE OUTER SURFACE OF THE LEFT HEMISPHERE OF THE BRAIN.

THE LOBES.

F.R.—Frontal lobe.

P.A.—Parietal lobe, consisting of (S.P.L.) superior parietal and (S.M., A.G.) infero-parietal lobules. T.S. Temperosphenoidal. O.C. Occipital lobe.

The Central lobe lies within the Sylvian fissure, and is covered by the posterior portion of the *Infero-Frontal* convolution, and the superior anterior part of the *superior* tempero-sphenoid convolution, where the fissure of Sylvius forms a figure like the letter Y. The cerebral mass that covers the central lobe is called the Operculum.

FISSURES AND SULCI.

F.S.—Fissure of Sylvius with its two branches.

H.S.—Horizontal branch of the Fissure of Sylvius.

A.S.—Ascending Branch of the Fissure of Sylvius.

R.—Fissure of Rolando.

O.C.—Parieto-occipital fissure.

I.P.—Intra-parietal fissure.

d.—Antero-ascending sulcus.

f.—Supero-frontal sulcus.

g.—Infero-frontal sulcus.

H.—The parallel fissure or superior tempero-sphenoidal sulcus.

i.—Inferior tempero-sphenoidal sulcus.

CONVOLUTIONS.

F.A.—Fronto-ascending convolution.

S.F.—Supero-frontal, M.F. medio-frontal, and I.F. inferofrontal convolutions.

P.A.—Parieto-ascending convolution.

s.m.—Supra-marginal or parietal eminence gyrus.

a.g.—Angular gyrus.

s.t.s.—Superior, m.t.s. middle, and i.t.s. inferior tempero-sphenoidal convolutions.

1, 2, 3, First, second and third ; or upper, middle and lower occipital convolutions.

PLATE II.

MEDIAN SURFACE OF THE RIGHT HEMISPHERE OF THE BRAIN.—*(After Quain.)*

c. Anterior fold (cornu) of the corpus callosum. d. Its posterior fold (splenium). g.f. Gyrus fornicatus. s.t. Supero-frontal convolution. a.p. Ascending frontal con-volution. p.p. Parieto-ascending convolution. q.l. Quad-rate lobule. q.u. The cuneus. c.b. Cerebellum. e. The fornix. f. The right crus of the fornix. g. The right peduncle of the pineal gland. h. The pineal gland. i. Section of the middle commissure of the third ventricle (commissura mollis). k. The iter à tertio ad quartum ventriculum. l. Section of the anterior commissure. m. Section of the posterior commissure. n. The septum lucidum. o. The tuber cinereum. p. The right corpus albicantium, showing the white fasciculus which it receives from the thalamus opticus. q. The pituitary gland. r. The divided edge of the valve of Vieussens. s. The fourth ventricle, t. The arbor vitæ. v. The fasciculus of the corpus pyramidal seen passing into the pons Varolii. w. The separation of the fibres of the corpus pyramidal, with the admixture of the grey substance : this is the motor tract (tractus motorius). x. The crus cerebri. y. The locus perforatus.

PLATE III.

THE BRAIN IN SITU.

Shows the situation of the brain in the skull, and the rela-tions of the one to the other. The right half of the

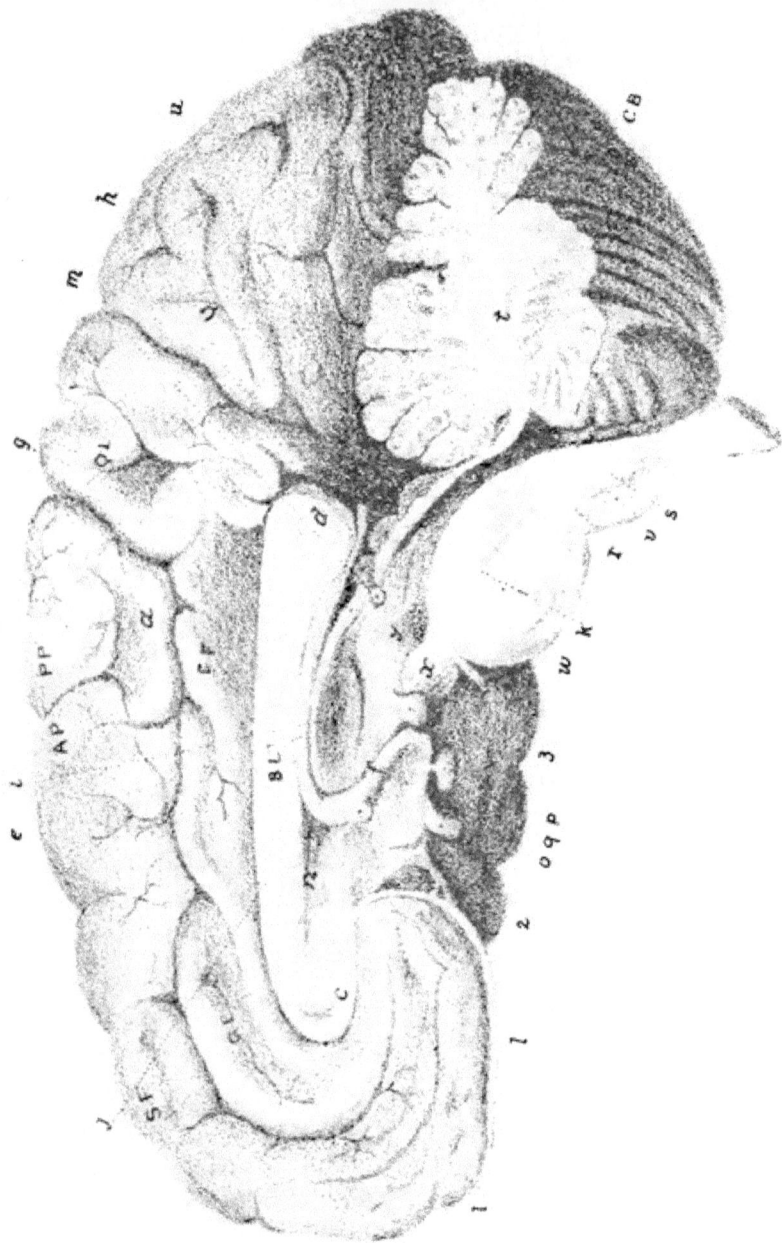

No. II

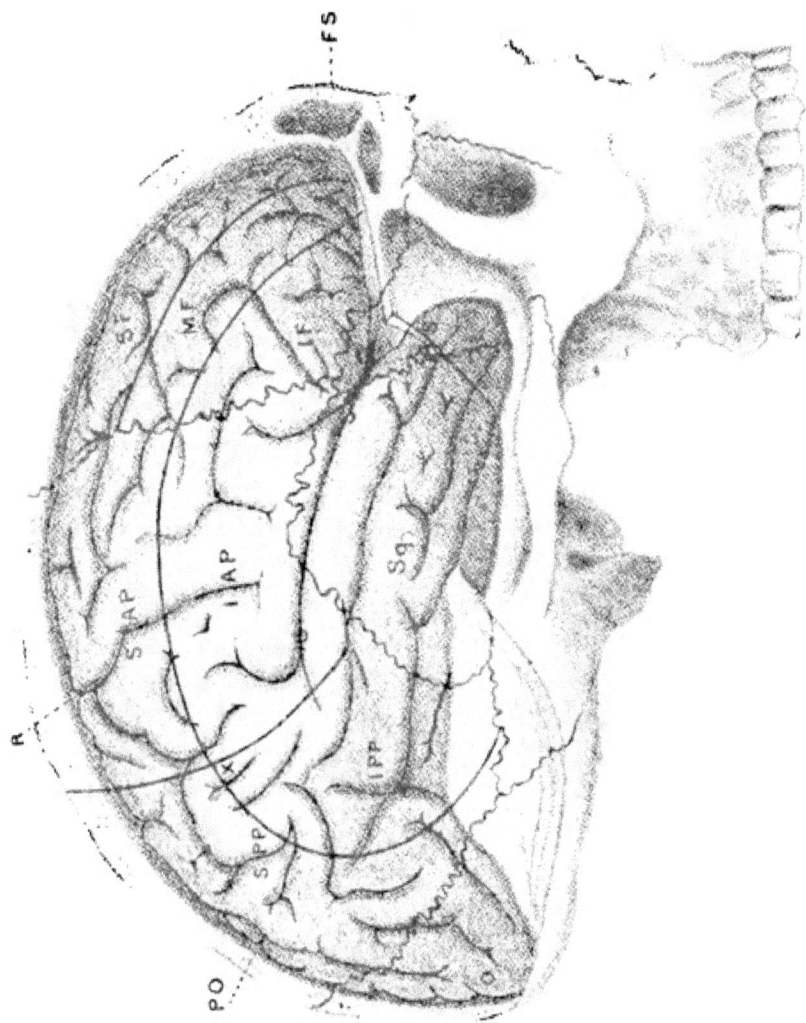

FS
ST
MF
IF
Sg
AP
SAP
IAP
R
SPP
IPP
PO

cranium is removed, exposing the outer cerebral convolutions to view. The positions of the sutures of the skull, the temporal ridge and parietal eminence (x) are engraved on the brain, also a short horizontal line parallel to the temporal ridge.

Professor Turner divides each hemisphere of the skull and brain into ten areas. Refer to pp. 116–29 for description, and to plate IV. for the situations of the sutures, etc.

S. Fissure of Sylvius. R. Fissure of Rolando. P.O. Parieto-occipital fissure. The frontal bone is divided into three areas, viz. :—supero-medio and infero-frontal areas. They are divided by the temporal ridge and an artificial line parallel to it, drawn from the centre of the orbit upward through the frontal eminence, and back to the coronal suture ; S.F., M.F., I.F., are drawn on the frontal convolutions to mark their situations. They are bounded behind and below by the coronal, fronto-sphenoid and parieto-sphenoid sutures. The *parietal bone* is divided into four areas,—horizontally by the temporal ridge, and vertically by a line drawn from the posterior superior portion of the squamous suture obliquely backward and upward so as to pass through the parietal eminence in its course to the saggital suture. The letters S.A.P., I.A.P., S.P.P., I.P.P., mark their situations. They are named supero and infero-antero-parietal areas : Supero-infero-postero-parietal areas. The *temporal bone* forms one area *(s.g.)*, named the squamoso-sphenoid. The wing of the sphenoid bone also forms one (A.S.), the ali-sphenoid area ; and each lateral half of the *occipital bone* (O) forms an area. The letters A.P. are engraved on the antero-ascending convolution, and S.I. on the parieto-ascending convolution. S.P.P. Are on the angular convolution, and I.P.P. on the mid-tempero-sphenoidal convolution, and *s.q.* are on the same. A.S. Are placed on the tip of the anterior portion of the supero-tempero-sphenoidal convolution. O. Is on the

mid-occipital convolution, and X. is on the convolution
of the parietal eminence, or supra-marginal gyrus.

PLATE IV.

RIGHT SIDE OF THE SKULL,

Showing the Cranio-Cerebral Map, after Professor Turner, with the
areas increased to seventeen by the Author.

A.—Frontal eminence, or point of ossification of the frontal
bone.

B.—Parietal eminence, or point of ossification of the parietal
bone.

C.—Squamoso-sphenoid suture.

D.—Temporal ridge.

E.—Superior curved line of the occipital bone.

F.—Occipital spinus process.

g.--Fronto-sphenoid suture.

h.—Parieto-sphenoid suture.

i.—Squamous suture.

K.—Coronal suture.

M.—Lambdoidal suture.

N.—Apex of the occipital bone.

P.O.—Situation of the parieto-occipital fissure of the brain,
and the parieto-occipital boundary line of the cranial areas.

R.—Dotted line indicating the position of the fissure of
Rolando.

S.—Transverse suture, and the external angular process.

v.—Inter-parietal boundary line.

X.—Anterior border of the glenoid fosse, in which the head of
the ramus of the lower jaw moves.

z.—Zygomatic arch.

AREAS.

s.f.—Supero-frontal area.

m.f.—Medio-frontal area.

i.f.—Infero-frontal area.

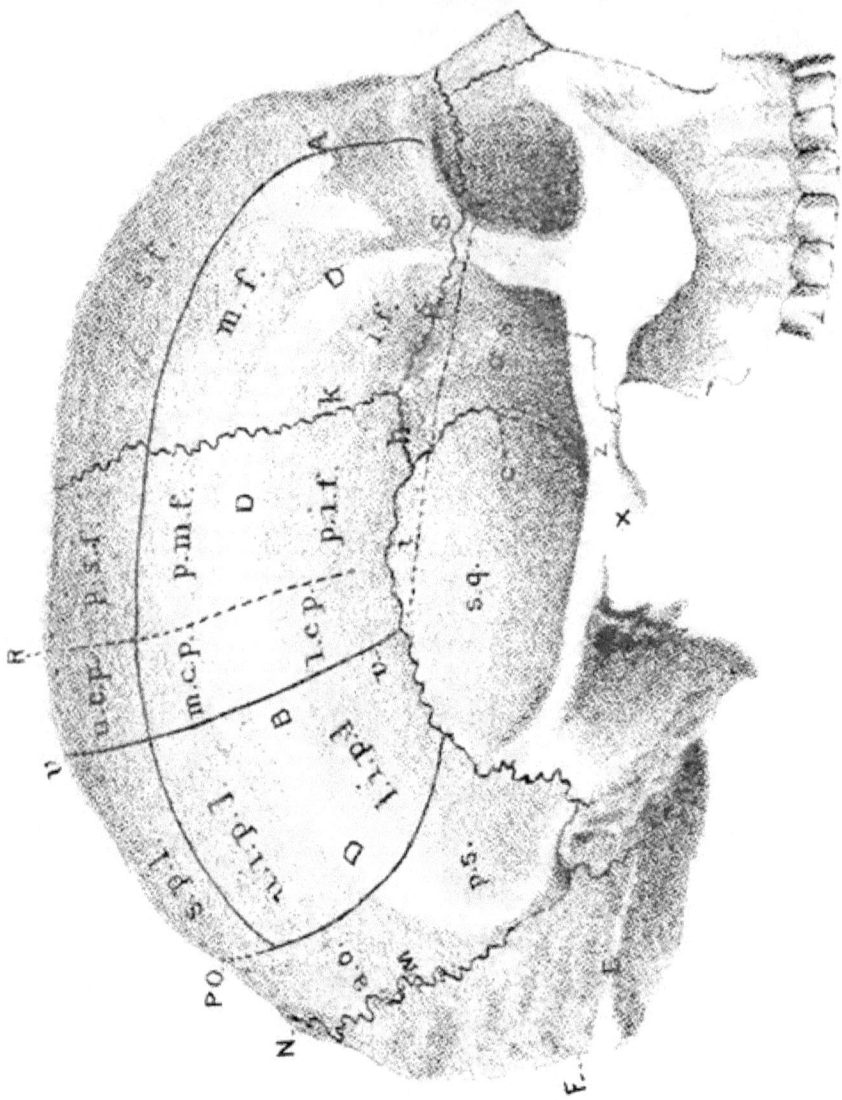

p.s.f.—Postero-supero-frontal area.

p.m.f.—Postero-medio-frontal area.

p.i.f.—Postero-infero-frontal area.

u.c.p.—Upper centro-parietal area.

m.c.p.—Middle centro-parietal area.

l.c.p.—Lower centro-parietal area.

.p.l.—Supero-parietal lobule area.

u.i.p.l.—Upper infero-parietal lobule area.

l.i.p.l.—Lower infero-parietal lobule area.

a.o.—Antero-occipital area.

N.E.F.—These letters and the lambdoidal suture bound the occipital area.

s.q.—Squamoso-sphenoid area.

a.s.—Ali-sphenoid area.

p.s.—Post-sphenoid area.

ILLUSTRATIONS.

b

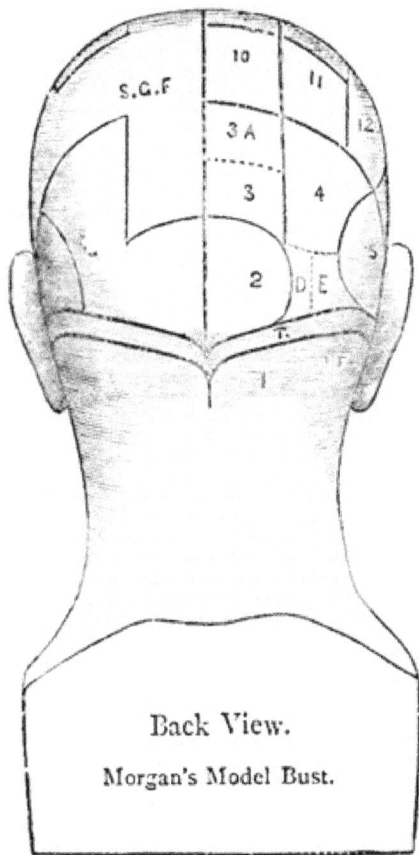

Back View.

Morgan's Model Bust.

8. Acquisitiveness.
9. Constructiveness.
10. Self-Esteem.
11. Approbativeness.
12. Cautiousness.
13. Benevolence.
14. Veneration.
15. Firmness.
16. Conscientiousness.
17. Hope.
18. Marvellousness.
19. Love of the Picturesque
19B. Sublimity.
20. Imitation.
21. Humorousness.
22. Individuality.
23. Form.
24. Size.
25. Weight.
26. Colour.
27. Locality.
28. Number.
29. Order.
30. Eventuality.
31. Time.
32. Tune.
33. Language.
34. Comparison.
35. Causality.
36. Graveness.
37. Gayness.
38. Awe.

1. Amativeness.
2. Philoprogenitiveness.
3. Inhabitiveness.
3A. Continuitiveness.
4. Adhesiveness.
5. Defensiveness.
E. The Centre of Energy.
6. Destructiveness.
6A. Alimentiveness.
C. Bibativeness.
7. Secretiveness.

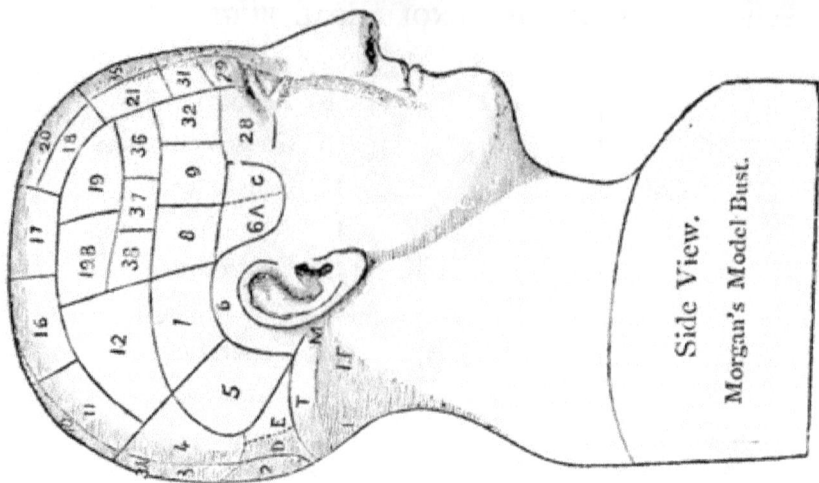

Side View.
Morgan's Model Bust.

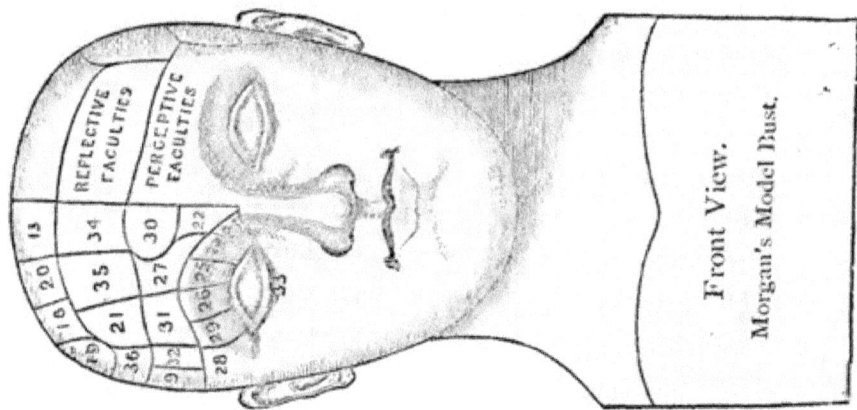

Top View.
Morgan's Model Bust.

Front View.
Morgan's Model Bust.

MORAL AND RELIGIOUS FACULTIES

REFLECTIVE FACULTIES

PERCEPTIVE FACULTIES

ERRATA.

For *tract.* p. 23, l. 4, and p. 30, l. 8, read "track."

For *vent*, p. 27, l. 10, read "event."

For *super orbital*, p. 72, l. 26, read "supra-orbital."

For *one late*,—the last word of p. 32, and the first word of p. 33,—read "articulate."

For *differential*, p. 78, l. 9, read "deferential."

CHAPTER I.

TO STUDENTS.

OBJECTIONS TO PHRENOLOGY.

I deem it advisable at the outset to briefly notice a few questions and objections which occasionally crop up in the minds of students as stumbling-blocks at the very threshold of inquiry, although they hardly come within the scope of an elementary treatise.

The first question which suggests itself is, Is phrenology true? I believe, in the main, it is: and that as a system for interpreting character it has no rival, notwithstanding it is comparatively young, and may require many years and much thought to mature it.

My design in this work is not so much to prove that phrenology is true, as to explain its principles and rules for the student's guidance. so as he may test the truthfulness of it for himself. I do not wish

any person to believe phrenology true on my dictum. To all I say,—Judge for yourselves by an
appeal to the facts hereafter stated, and to Nature.
Do not, however, be in a hurry in forming a judgment. Be patient and deliberate in your investigations. Master the principles and rules, and
apply them thoughtfully and impartially, leaving
your minds open to conviction. Be not over-
credulous. I should rather you were a little sceptical to begin with, if you be inspired by thorough-
going earnestness to arrive at the truth, and a determination not to be deterred by obstacles in its
pursuit. If this be the student's aim, and he have
sufficient natural aptitude, no fear need be enter-
tained of the result.

In one or two reviews of *Phrenology and How
to Use it in Analyzing Character*, my attention was
directed to the objections brought against phrenology by Professor Bain on the psychological side,
and Mr. G. H. Lewes on the physiological side of
the subject, as being authorities that cannot properly be silently overlooked; an opinion in which
I concur.

Professor Bain has given the most temperate,
and apparently impartial criticism on phrenology,
—in *Study of Character*,—that has come under my
notice, which I had read with pleasure and profit
before writing *Phrenology, and How to Use it in
Analyzing Character*, but he treats almost exclu-
sively on the psychological bearings of the phreno-

logical analysis, and, as such, it does not come within the province of the present work.

Mr. Bain, however, emphatically states that, as a science of character, phrenology has no rival, although he takes many exceptions to it as a psychology, and points out clearly what he considers its errors and omissions in this department, but he produces no facts of want of correspondence between cranial conformation and mental tendencies in accordance with the phrenological doctrines.

A very different course is pursued by Mr. Lewes, and his criticism demands a more extended notice. He makes direct charges against the doctrines of practical phrenology, or craniology, as he calls it, and brings forward several cases as crucial tests to show the want of agreement between cranial development and intellectual capacity, in an article in the December number of Blackwood's *Magazine*, A.D. 1857, entitled "Phrenology in France," and he repeats the substance of it without modification in the latest edition of his "History of Philosophy," published in 1871. We may, therefore, fairly assume that the arguments it contains are the strongest the author can advance.

In Blackwood's *Magazine*, Mr. Lewes says that, though "Phrenology was born in Germany and reared in France, it has not standing room in either country." The chief reason of this state of things, according to Mr. Lewes, is, that "the persons who interest themselves at all in psychology are invari-

ably either psychologists or physiologists." But in
England the case is different,—" besides these
classes there is a very large class of what may be
loosely styled the general thinking public," who are
" willing to give all systems a fair hearing ; consequently, phrenology not only has standing room in
these countries, but has grown to bulky dimensions."
This is highly suggestive. The seed of phrenology
germinates and grows vigorously in these countries,
because it is sown in the soil of thought, and is
manured by impartial investigation, and warmed
by the sunshine of sincerity. No wonder, then,
that it withstands the blight of ridicule and the
storms of prejudice. Mr. Lewes declares, "We have
not the slightest intention of discussing the principles of phrenology in this article. . . . Our
purpose is restricted to the bringing under the
reader's notice some rather startling objections to
the phrenological doctrines which have been advanced by M. Louis Piesse, in a series of articles
just published by him, in his two agreeable and
suggestive volumes of medical essays."—*La Médecine
et les Médecins*, etc. Paris 1857.

Mr. Lewes tells us, that " M. Piesse does not examine the principles of phrenology. He simply
takes a few striking examples as they arise, and examines how far they are in accordance with phrenological teaching." Why, Mr. Lewes, are the examples
so numerous, so overwhelming in force that M. Piesse
need not trouble himself by making a selection ?

This is the natural inference. How far this is borne out by fact the sequel will show. Sir John Forbes, in the *British and Foreign Medical Review*, remarks, " That anti-phrenologists were bound to collect an equal mass of contradictory facts, if they seriously meant to refute the doctrines of phrenology." To this Mr. Lewes replies :—" The trouble was great, the motive was not sufficiently strong !" Anti-phrenologists have condemned the subject, that is enough. " Besides," says he, "ordinary observation supplied them with instances of failure too unequivocal to permit their acceptance of phrenology, and they left to others the onerous burthen of collecting an imposing array of facts or counter-evidence." This paragraph is singularly contradictory, and more especially so when taken in connection with the fact as assumed by Mr. Lewes of M. Piesse not needing to trouble himself much in the collection of striking examples. First, we are told the trouble was great, and the motive too weak, to induce anti-phrenologists to follow the advice of Sir John Forbes ; and second, that ordinary observation supplied numerous instances of failure : Mr. Lewes continues, " M. Piesse has supplied us with no great array of facts, but those selected by him are sufficiently significant, and make up in quality for their deficiency in quantity." Here M. Piesse is said to have *selected* significant facts, but in the preceding paragraph he is said to have taken them as *they arose*. Those selected facts appear

to have overwhelmed Mr. Lewes's judgment. " Indeed," he says, "after reading his (M. Piesse's)
account of Napoleon, Feischi, Lacenaire, the calculating boy, and the young Indian girl, we can feel
no surprise at phrenology being discredited in
France." Mr. Lewes, in another place, charges
phrenologists with a want of scientific scepticism,
and with being led off the tract of scientific inquiry by the influence of bias. We, therefore,
naturally expect better things of him : that he will
avoid the short-comings of phrenologists, and investigate the evidence for and against the subject
impartially, and with all the minuteness of a scientific
sceptic, before coming to a decision. Let us see,
then, how he demeans himself in the examination
of M. Piesse's so-called facts. He actually receives
them as being indisputably true from the pen of M.
Piesse alone. He never questions their correctness
—never for a moment thinks that M. Piesse might
have been biased, and thereby led astray from the
tract of true scientific inquiry. Surely, Mr. Lewes
was bound to have taken some pains to verify M.
Piesse's statements, before he brought them forth so
prominently for the determination of such weighty
issues, otherwise he should have left the responsibility resting on M. Piesse, and allowed him to
support his own case ; but this would not have
served Mr. Lewes's purpose. He seems to have
been waiting for a favourable opportunity to strike
a death-blow at phrenology, and under the covert

design of reviewing M. Piesse's works, he tries to
hit his aim ; but I am anticipating. Let me now
draw attention to M. Piesse's facts which are said
to be so significant and so excellent in quality.
Mangiamele, the son of a Sicilian shepherd, when
a boy, astonished the Académie des Sciences by
his marvellous powers of calculation. At this time
he was only eleven years of age, and was entirely
self-taught. Yet the rapidity with which he solved
the most intricate arithmetical problems without the
aid of graphic signs was marvellous. "This," observes Mr. Lewes, "was a crucial test for phrenology. . . . And the reader will learn with
interest that M. Piesse pointed to the skull of this
boy, and showed a decided *depression* at that particular spot where the organ of Number was situated
—a depression instead of an eminence. The reader
will learn with increasing interest that this contradiction between theory and fact was confirmed by
the phrenologists themselves ; for, instead of deny-
ing the depression, they endeavoured, as usual, to
explain it." Now, only two phrenologists, Broussais
and Dumortier, are said to have undertaken to
explain this contradiction between theory and fact.
Yet these two are spoken of as the phrenologists,
as though the whole of the phrenologists of France
had done so. They are represented as having admitted that there was no special development of
the faculty of Number indicated in Mangiamele's
skull ; and that they accounted for his talent for

calculation by the activity of other faculties,—such
as Causality, Comparison, Individuality and Eventuality.   If it be true that Broussais and Dumortier
attributed this boy's power of calculation to other
faculties than Number—in the absence of the usual
indication—they did a very foolish thing, and de-
servedly brought down ridicule on themselves, and
indirectly on the system they wished to defend ;
and if there was actually a depression in the part
of Mangiamele's head at the organ of Number, it
certainly was a strong fact against the localization
of this organ.   I say, if these things were as represented by M. Piesse : I am, however, somewhat
sceptical on this point.   My reason for doubting
the correctness of M. Piesse's report will shortly
transpire.

After Mr. Lewes has made the best of Mangiamele's case to crush phrenology, he endeavours to
strengthen his position by citing a few more cases
as damaging facts.   First in the list is Napoleon I.
He says, " Let us with M. Piesse examine the case
presented by Napoleon.   A few hours after his
death, a cast of Napoleon's face and the anterior
half of his skull was taken by Dr. Antomarchi.
.   .   .   That cast comprises the greater number
of the phrenological organs, and all those of the
higher faculties."   " Artists," continues Mr. Lewes,
" have grossly exaggerated the head of Napoleon.
But what says the cast ?   The head is decidedly a
small one.   .   .   It is, however, extremely

well proportioned. Its circumference being twenty inches ten lines (French measurement), its dimensions by no means remarkable."

Mr. Lewes, in a foot note, makes the following *naïve* remarks on M. Piesse's mode of measurement and results : " M. Piesse has not explained how he arrived at this precise measurement in the absence of the back part of the skull ; but, from the specific size given, we presume he had some positive data." This is a remarkable presumption. How could M. Piesse have had " positive data" to supply the back head of Napoleon, so as to measure the circumference of the head precisely ? It was impossible. How Mr. Lewes could fancy by any stretch of his imagination that M. Piesse had had positive data I cannot divine, excepting on the supposition that his anti-phrenological bias had dimmed his understanding ; inasmuch as he knew this was the only cast taken of Napoleon, for he remarks : —"We learn from M. Piesse that some phrenologists disavowed the fidelity of the cast, taken as it was by one not expert in the art. Faithful or not faithful, it is the only authentic cast which exists."

Mr. Lewes continues, "we must interrupt our analysis of M. Piesse's essay to observe that Mr. Combe (and we presume other phrenologists) states, as if it were a simple matter of fact, that Napoleon's head was large ; and, indeed he cites Napoleon as an example between correspondence of size of head and force of character. Not having taken the mea-

sure of the skull, we cannot pretend to judge which
of these two statements is correct." Notwithstanding Mr. Lewes had not measured the cast of Napoleon, he throughout this article assumes that M.
Piesse's measurements are correct, and that Mr.
Combe and other phrenologists simply take it for
granted that Napoleon's head was large. Now, Dr.
Antomarchi's cast of Napoleon was in the possession
of the phrenological society of Edinburgh, and Mr.
Combe had measured it, before citing it as an
example of large size. Mr. Lewes would lead his
readers to believe, though he does not say so, that
Mr. Combe and other phrenologists had neither
the disposition nor the means of testing the accuracy
of M. Piesse's statements, but that they simply
assumed them to be wrong, because they told against
their doctrines. If phrenologists had followed this
course they would have only done what Mr. Lewes
himself did. He admits not having measured the
cast, but he speaks of it as a decidedly strong contradictory fact against phrenology, and he enters
into particulars in order to make out his case. He
says, "The faculty of Ideality or imagination, for
example, was certainly large in Napoleon ; but the
head shows no remarkable development in the region marked as that of Ideality ; while the organ of
Number has a strongly marked depression. A depression in the region which embraces Number,
Constructiveness, and Order is surprising, M. Piesse
observes, in Napoleon, so remarkable for his mathe-

matical tendencies.   We should add, however, that neither Spurzheim nor Mr. Combe considers Number to be the mathematical organ,—they limit it to arithmetic. Still phrenology is called upon to show that, with this depression of Number, Order, and Constructiveness, in the skull of Napoleon, there was a corresponding absence of those faculties in his intellect;" but "few will have the hardihood to make such assertions." We are also told by Mr. Lewes, that "the organs of Causality and Comparison were not more developed in the skull of Napoleon than they are in ordinary men. M. Piesse has measured the frontal angle, and finds it 75°.   In truth, the forehead of Napoleon, phrenologically speaking, was quite mediocre, of which any candid observer may convince himself. Simple inspection suffices to assure us of this fact, and the measurement 75° is a geometrical proof. M. Piesse sums up his investigation with remarking that the veritable head of Napoleon, studied according to phrenological principles, so far from confirming Gall's doctrine, completely refutes it."

Before proceeding further, let us examine the case of Napoleon. I regret very much not being able to test the truthfulness of M. Piesse's narrative of Mangiamele, by an examination of a cast of this boy's head. I am inclined to think one was taken, and shall never rest satisfied until a copy of it is in my cabinet, if one can be got. I have already caused inquiries to be made in Paris respecting it.

Happily, fortune favours me in the case of Napoleon. An authentic copy of Dr. Antomarchi's cast of him, is on the table before me. No person who has been accustomed to examine living heads, casts from nature, or skulls, could possibly fall into the error of supposing the head of Napoleon was small by looking at this cast. Every unprejudiced person of ordinary discrimination would unhesitatingly arrive at a very opposite conclusion ; and the very fact of M. Piesse having misrepresented Napoleon's mask, as I now proceed to prove he has done, shows him to be an unreliable witness. In fact, the measurements of the cast so completely disprove M. Piesse's testimony in this case that, without irrefragable, corroborative evidence, his statements respecting Mangiamele's organ of Number must be discredited. Although a portion of the back-head of Napoleon is not shown in the cast, there is strong reason for inferring that his head was considerably above the average size of the European male head. His intellect was large, according to phrenological principles ; his reflective organs were largely developed, especially Comparison ; the organs of Order and Ideality were well developed, and the organs of Number and Constructiveness were full, or rather larger than what they are in the average run of persons.

That Napoleon's head was large is proved beyond any possibility of a doubt by comparing it with the heads of Samuel Taylor Coleridge and Dr. A.

Combe. Coleridge had a very large head and a powerful intellect, as is generally well known. Dr. A. Combe, likewise, had a large head, and talents of a high order. Now, I have measured the anterior portion of the casts of these heads from the external opening of the ear, and Napoleon's by the same method, and find that the latter is the largest, as will be seen by the following tabulated measurements.

Name.	Inches.				Inches and tenths.						
	a	*b*	*c*	*d*	*e*	*f*	*g*	*h*	*i*	*j*	*k*
Coleridge	14⅛	12½	11	6½	4·4	3·7	4·5	5·2	5·5	5·6	6·3
A. Combe	14¼	12⅜	11¾	7	4·1	4·0	4·4	5·0	5·3	5·3	6·0
Napoleon	14⅞	13⅛	12¾	7	4·8	4·5	4·5	5·1	5·4	5·5	6·3

DESCRIPTION OF THE MODES OF MEASUREMENT.

(*a*).—This is a line drawn perpendicularly from the centre of the opening of one ear over the crown to the same point in the other ear, and forms a base or point of starting and ending to the lines registered in the columns *b* and *c*, and a starting point only to the lines *d*, *e*, and *f*.

(*b*).—Length of a horizontal line drawn from *a* at one side, over the superciliary ridge to *a* on the other side.

(*c*).—Length of a line parallel to *b*, drawn from the same point over the points of ossification of the frontal bone, or organ of Causality.

(*d*).—Length of a line drawn from *a*, at the crown, along the medial line to the root of the nose.

(*e*).—Length of the base of the forehead from *a* to the superciliary ridge.

(ƒ).—Length of the upper part of the forehead from *a* to the point of ossification of the frontal bone.

(*g*).—Breadth of the head at the organ of Order.

(*h*).—Breadth of the head at the organ of Number.

(*i*).—Breadth of the head at the organ of Constructiveness.

(*j*).—Breadth of the head at the organ of Ideality.

(*k*).—Breadth of the head at the organ of Secretiveness.

Napoleon's head before the ears is larger than either Coleridge's or Dr. Combe's. When we consider the large size of the head of Coleridge and of Dr. A. Combe, the former being 23½ inches in the horizontal circumference at the base, and the latter 23¼ inches, we could not be wrong by inferring that the head of Napoleon would be equally large. The cast of his head along the medial line extends two inches behind the opening of the ear, and indicates that the backhead was proportionally large with the frontal region, in accordance with the male type. However, be this as it may, that Napoleon's head was large is quite certain ; and that he had a massive intellectual organism, can admit of no doubt. The length of the forehead at the upper part is only three-tenths shorter than the lower part, whereas the difference in Coleridge in this respect is seven-tenths, and in Dr. Combe one-tenth. The analytical and mathematical organs of Napoleon are very large. His organs of Number, Constructiveness, and Ideality are only one-tenth each smaller than those of Coleridge, or the head of the former is only one-tenth narrower at the seats of these

organs than the latter, and Napoleon's organ of
Ideality is two-tenths larger than Dr. Combe's. The
narrowest part of Napoleon's head is at the seats of
the organs of Number and Constructiveness ; but
there is no depression at these parts.   This will be
clearly understood from the fact that the head, at
the organ of Number, is six-tenths wider than at
the organ of Order.   Let people measure their own
heads at the organs of Number and Constructiveness, and they will see that, by comparison, Napoleon's organs of these faculties are by no means
small.   Napoleon's organ of Secretiveness is large.
His head at this part being the widest,—namely,
6·3 inches.   The activity of this faculty, combined
with his splendid mathematical and analytical
talents, account for his strategetic skill.   The coronal region is large.   The cast of Napoleon's head,
viewed from the front, indicates much more than
ordinary vigour and mental acuteness, and when
seen in profile, depth of penetration, and comprehensiveness of understanding, are markedly portrayed.

Let us now examine the frontal angle which
Mr. Lewes, on the authority of M. Piesse, says was
only 75° in the skull of Napoleon, and which, he
advances as a geometrical proof, that his forehead,
phrenologically speaking, was only mediocre.   I remark first, that the preceding measurements, the
correctness of which, can easily be tested by Mr.
Lewes if he be so minded, irrefragably prove that
the head of Napoleon, instead of being mediocre,

was considerably larger than the average European
male head, and demonstrate the uselessness of
Camper's angle as a measure of comparative intellectual capacity.

The following description of the frontal angle,
with remarks on it by Blumenbach, is quoted from
his "Anthropological Treatises" as translated and
edited by T. Bendyshe, M.A., Fellow of King's
College, Cambridge, 1865, pp. 235–6 :—

59.   FACIAL LINE OF CAMPER.

" He imagined, on placing a skull in profile, two
right lines intersecting each other.   The first was
to be a horizontal line drawn through the external
auditory meatus and the bottom of the nostril.   The
second was to touch that part of the frontal bone
above the nose, and then to be produced to the
extreme alveolar limbus of the upper jaw.   By the
angle which the intersection of these two lines
would make, this distinguished man thought that
he could determine the difference of skulls as well
in brute animals as in the different nations of mankind."

60.   REMARKS UPON IT.

" But, if I am correct, this rule contains more
than one error.   *First*, what indeed is plain from
those varieties of the racial face I was speaking
of (s. 56), this universal facial line at the best

can only be adapted to those varieties of mankind which differ from each other in the direction of the jaws, but by no means to those who, in exactly the contrary way, are more remarkable for their lateral differences.

"*Secondly*, it very often happens that the skulls of the most different nations, who are separated, as they say, by the whole heaven from one another, have still one and the same direction of the facial line ; and, on the other hand, many skulls of one and the same race, agreeing entirely with a common disposition, have a facial line as different as possible.   We can form but a poor opinion of skulls when seen in profile alone, unless at the same time account be taken by their breadth.   Thus, as I now write, I have before me a pair of skulls,—viz., an Ethiopian of Congo, and a Lithuanian of Sarmatia. Both have almost exactly the same facial line ; yet their construction is as different as possible if you compare the narrow, and, as it were, keeled head of the Ethiopian with the square head of the Sarmatian. On the other hand I have two Ethiopian skulls in my possession differing in the most astonishing manner from each other as to their facial line, yet in both (if looked at in front, the narrow end, as it were, squeezed-up skulls) the compressed forehead, etc., sufficiently testify to their Ethiopian origin.

"*Thirdly*, and finally, Camper himself, in the plates appended to his work, has made such an arbitrary
c

and uncertain use of his two nominal lines, has so often varied the points of contact according to which he has drawn them, and upon which all their value and trustworthiness depend, as to make a tacit confession that he himself is uncertain, and hesitates in the application of them."

If any more proof is wanting of the uselessness of this angle for the purpose to which Mr. Lewes applies it, we have it in the fact that it was only 72° in Mr. George Combe, whose forehead no person will have the hardihood to say was mediocre ; that it is 75° in Dr. Livingstone, as shown by a photograph of him by Mayall, and is 77° in Canon Kingsley, whose forehead is prodigious ; and that it was 83° in the head of Dr. Pritchard, the Glasgow poisoner, whose forehead was comparatively small ; by which it will be seen, that 75° is no proof of a mediocre forehead.   The fact of Mr. Lewes citing 75° as a geometrical proof shows he has not paid sufficient attention to the subject to speak authoritatively on it.   Much more might be advanced on this head, but I leave the facts already stated to speak for themselves.

M. Piesse is represented by Mr. Lewes as saying that the heads of Descartes, Voltaire, Raphael, Fiechi, Lacenare, Avril, as well as Napoleon's, were small.   But after finding that M. Piesse has so misrepresented the case of Napoleon, I cannot confide in any statement of his that cannot be brought to the crucial test of matter-of-fact.

Mr. Lewes repeatedly states that the contradictory cases against the doctrines of phrenology are very numerous ; and that ordinary observation supplies sufficient evidence to upset the hypothesis. If this be so, it would have been easy for Mr. Lewes to point to a few notable examples in this country that could be tested ; but he has not done so. We are, therefore, left to surmise the reason of his non-reference to British cases. Was it that "the trouble was great and the motive insufficient for the task ?"

Mr. Lewes reviews Gall's system at considerable length in " History of Philosophy," fourth edition, vol. ii., and pays him a high tribute for his contributions to science, and the immense service rendered to philosophy and psychology, and says :—

"He may be said to have definitely settled the dispute between the partisans of innate ideas and the partisans of sensationalism by establishing the connate tendencies, both affective and intellectual, which belong to the organic structure of man.—p. 417. . . . In his vision of psychology as a branch of biology, subject therefore to all biological laws, and to be pursued on biological methods, Gall may be said to have given the science its basis."—p. 423.

Mr. Lewes, however, takes exception to Gall's method of systematizing his discoveries. and strongly urges the non-verification of his hypothesis,—that is, that the cerebral hemispheres consist of a congeries of organs, each being the organ of a distinct mental faculty :—

"The convolutions of the brain, which Gall has mapped out into several distinct compartments, each compartment being the organ of a distinct faculty, are in reality not more distinct than several folds of a piece of velvet; and a little more reflection discloses the absurdity of supposing that one portion of this velvet could be endowed with different properties from every other portion, simply in virtue of its superficial position. The tissue of which the convolutions consist is the same throughout its folds."—*Ibid*, p. 432.

This illustration looks somewhat formidable in point of argument at first sight, but on closer inspection its fallaciousness is seen to stand prominently out.   If similarity of cerebral tissue is proof of oneness of function, it would be absurd to infer that the anterior and posterior lobes are differentiated.   Yet this is the universal opinion. Medico-psychologists are agreed on the point of the anterior lobe of the brain being the seat of the intellect; in other words, both psychologists and medico-psychologists are *regional* phrenologists. But Gall has nowhere said, as Mr. Lewes leads his readers to infer, that the convolutions are different in function, simply in virtue of their superficial position, nor has any of his followers propounded such an absurdity.   Nor did Gall map out the convolutions into distinct compartments as organs of distinct faculties.   He has stated over and over again that he simply observed a general correspondence between particular prominences of persons' heads and their mental manifestations, and was led to infer that the cerebral masses immediately underlying such prominences have special functions, and that he investigated the structure of the brain in search of confirmation of this view, and saw, therein, nothing at variance with it, but much in corroboration thereof; nay, in fact, he thought he had obtained convincing evidence of there being such a relation. So far, then, from Gall having mapped the convolutions into distinct organs, he only observed and made known Nature's revelations. How far he succeeded in this laudable desire, time will reveal. In order that no mistake might be made regarding his observations and deductions, he has stated the facts and explained his procedure with great plainness and minuteness in his works; furthermore, he desires all who wish self-conviction to appeal to Nature, as "*there is no other irrefragable authority.*" Gall did not content himself with observing the connection between the form of the head and mental manifestation. He summoned pathology to his aid, and lost no opportunity of investigating the effects of disease of the brain on the operations of the mind, and of studying the connection between local cerebral lesion and specific mental aberration.

Mr. Lewes (p. 451) says :—

"The subject of the convolutions is one which might furnish an instructive chapter, did space permit, but I must content myself with affirming that the researches of anatomists have disproved every point advanced by Gall."

I shall content myself, in reply to this affirmation, by pitting against it the researches in cerebral physiology of Dr. Ferrier, who has given considerable attention to this subject. The doctor, following the course struck out by Fritsch and Hitzig, to whom he gives the credit of being the first to demonstrate that the brain is not, as has been generally stated, insusceptible to every kind of irritation, has experimented on the exposed cerebral hemispheres of pigeons, fowls, guinea-pigs, rabbits, cats, dogs, and monkeys by Faradisation (electrical irritation), and succeeded in producing local irritation of various parts of the brain. His method is to narcotise the animal with chloroform and to extend it on a board with its abdomen downwards, secured by cords, so as to give the head and limbs free play, then to expose the brain by trephining, and to extend the orifice by bone forceps. The process of Faradising is next proceeded with. This is done by electrodes made of thin copper wire, doubled at the end, rather obtusely, or in a slightly rounded form, so as to avoid laceration of the parts to which they are applied. They are likewise covered by silk thread, excepting a small portion of the end, and the doctor applies them, about a quarter of an inch separate to the parts of the convolutions that he wishes to test as to whether or not they are different in function.

After much experimental research and repeated verification of the phenomena, Dr. Ferrier arrived at the conclusion that the *individual convolutions are separate and distinct organs.*

Dr. Ferrier's researches irrefragably prove that Dr. Gall was on the right tract, and they verify his hypothesis, so far as they go,—namely, that the cerebral hemispheres are not homogenial in function, but are constituted of a number of distinct organs, having distinct functions.   What will Mr. Lewes say after this?   He surely will deem it advisable in the next edition of "History of Philosophy" to modify his views, and to eliminate the gratuitous affirmation "that the researches of anatomists have disproved every point advanced by Gall," in the same way as he found it necessary to rewrite many portions of the last edition, so as to bring them more in consonance with the spirit of the times.

Dr. Ferrier's experiments are valuable, and calculated to push on mental-physiological investigation ; but his conclusions are not to go unchallenged. Already dissentient voices are being heard, both near and in the distance.

In a letter received July 25th, 1874, from Dr. W. A. F. Brown, Psychological Consultant, Crighton Royal Institution, and late Commissioner in Lunacy, referring to this subject, he says :—

"Dr. Ferrier's experiments, if they show anything, prove that the grey matter is connected with the motor expression of irritation,—or, what phrenologists believe to be more likely, that the electrical irritation *conducted* merely through

the grey matter, acted upon the white matter generally, or
upon the *corpora striata* in particular, thus producing the
phenomena described."

M. Carville and M. Dupuy, of Paris, have both
shown, by different methods, that weak induced
currents are capable of diffusion to a distance in
the cerebral substance; and they account for the
movements produced by Faradisation to excitement
of parts at a distance from those to which the electrodes are applied. Dr. Brown-Sequard, of New
York, also holds the same opinion.

Dr. Ferrier, however, it is almost needless to say,
disputes that the motor effects were produced either
by weak diffused currents, or by direct electrical
stimulation of the motor nerves; and the following
facts may be cited in support of this opinion :—
(1), That Faradising the cortical substance of the
cerebrum increased the circulation in it to a preter-
natural flow, or an hyperæmic condition; and that
the motor effects were strongest when the hyperæmia was greatest; and further, the hyperæmia
seemed to increase in a ratio with the increased
distance between the electrodes, or the farther they
were applied apart on the convolutions; and (2),
that no motor effect was produced by Faradising the
cerebral medullary substance.

These facts show that the electrical current
either did not pass through the grey matter to the
white, as supposed, or that the white matter is not
susceptible to electrical stimulation disconnected

from the grey; but the former view appears the more rational.

Whether or not Dr. Ferrier's conclusion regarding the excitability of the grey matter be correct, his experiments show differentiation of function, inasmuch as very different effects were produced by the application of the electrodes on different points of the convolutions. For, supposing the electrical current did not act directly on the grey matter, but passed on to the white, the various fibres of this substance must have different connections, in order to effect such diverse movements as described and attested by numerous scientific witnesses.

In awarding to Dr. Ferrier the praise he merits, we must be careful not to detract from what is due to his predecessors, who have traversed the same field of inquiry by different routes, and who demonstrated the differentiation of function of the hemispheric ganglion many years ago. He has a vast territory to explore before he overtakes some of them. The utmost he has accomplished is the mapping out of a few motor centres. The centres, the functions of which are to give expression to the phenomena of thought, have eluded his search, and I am inclined to think they will continue irresponsive to Faradisation. "How shall we," asks the *Scotsman*, "by such means fathom the intellectual and moral life of man? How shall we by such crude experiments make manifest the existence of an intellect that is capable of tracing the action of

gravitation throughout the boundless regions of
space; or trace the cause or origin of those moral
feelings which make up so much of the sweet and
bitter of human life."

The small amount of evidence that serves some
persons to prop up particular theories is remarkable.
Dr. Ferrier having produced movements of the eyes
of animals by Faradising the cerebellum, and failing
to produce any amative symptoms, the conclusion
that it has no connection with the generative instinct
is come to forthwith.*

The mass of evidence, both positive and negative,
produced by Gall (" Functions of the Cerebellum"),
showing a general concomitance between the disposition to functional action of the amative propensity
and the size of the cerebellum is ignored, and the
doctrine that it is the centre of the instinct of generation is pronounced untenable.

When the emotions and the intellect of man are
found susceptible to being evolved by Faradisation, and the cerebellum has been thoroughly ex-
plored by the same means without evoking any
manifestation of sexual love, then we may reason-
ably doubt the phrenological doctrine regarding its
functions; but, until this is accomplished, and so
long as Gall's evidence remains intact, we shall be
justified in following his teaching on the subject.

At the time Dr. Ferrier's researches in cerebral
physiology were made known, some newspaper cor-

* Mental Physiology.

respondents predicted, with a jubilant air, that his discoveries would not be comforting to phrenologists !
Why not ? Was the wish the parent of this predic-
tion ? Are phrenologists so wedded to their belief
as to bolster it up at the expense of truth ? I
answer for myself, and declare my readiness to give
up any,—nay, every position held by phrenologists,
as soon as convincing proof against them shall be
produced ; and, moreover, should, as a lover of
truth, hail the event with joyous acclamations ; and,
I believe, phrenologists generally would join in the
chorus.

Referring to Dr. Ferrier's experiments, Dr. Carpenter says :—

"We have to enquire how far these experimental results
justify the belief that there is any such localization of strictly
mental states, as there is of the centres for the expressing of
those states in movement. And, as to this, it must be con-
fessed that we are still very much in the dark. . . . .
Although there would seem strong ground for the belief that
the memory of particular classes of ideas *may* be thus localized,
. .· . and that particular parts of the convolutions may be
special centres of the classes of perceptional ideas that are
automatically called up by sense-impressions."—*Mental Physiology* (Appendix), p. 721.

These are notable facts. They conclusively
point out the transitional state which is taking place
in the minds of medico-psychologists regarding the
cerebral functions.

Dr. Ecker, Professor of Anatomy and Comparative Anatomy in the University of Freiburg, Baden, begins the introduction to his work on " The Anatomy of the Human Brain," by the following noteworthy passage, which, although it has no direct bearing on Ferrier's researches, is sufficiently significant of the signs of the times, as to justify its reproduction here :—

" That the cortex of the cerebrum, the undoubted material substratum of our mental operations, is not a single organ, which is brought into play as a whole in the exercise of each and every psychical function, but consists rather of a multitude of mental organs, each of which is subservient to certain intellectual processes, is a conviction which forces itself upon us almost with the necessity of a claim of reason. The hypothesis set up in opposition to it, of a single organ for the carrying out of the multiplicity of psychical functions, would present about an equivalent point of view to that of ' vital force,' which has received its *coup de grâce*. If, however, as we conceive to be an undoubted fact, certain portions of the cortex of the cerebrum subserve certain intellectual processes, the possibility is at once conceded that we shall some day arrive at a complete organography of the surface of the brain —a science of the localization of the psychical functions."

What, I again ask, will Mr. Lewes say of this?

Since Dr. Ferrier's experiments on the animals previously referred to, he has experimented on monkeys, and produced, according to Dr Carpenter, far more remarkable results, which are of a more distinctly expressive character. But he seems to have discovered boundary lines, beyond which he is unable to pass. This has only now caught my

eye in turning over the pages of " Mental Physiology."

It appears that in those animals which have the
middle lobes of the brain fully developed, such
as the cat and dog, "stimulation of the posterior
portions of these lobes produces no respondent
movement." This is the case also in the monkey ;
but not only so, "*the whole of the posterior lobe is
similarly irresponsive*, as is also that *front* portion of
the *anterior* lobes," which in all the higher mam-
malia, as in man, is shown by that forward as well as
lateral development, which markedly distinguishes
it from the corresponding part of the cerebrum of
the rabbit. " What," significantly remarks Dr. Carpenter, "may be the special functions of these parts
we can scarcely do more than guess at." Guess at !
Are there no other means of arriving at the meaning of these significant facts than guessing, doctor?
Did not Gall interpret some of Nature's meanings
without Faradisation,—meanings that were called
absurdities, and held to be such. But now there is
obviously a shaking amongst the dry-bones of un-
belief. Why not give Gall's method as patient,
impartial, and persistent a trial as has been given to
Broca's theory of aphasia, and to Ferrier's researches?
Were this course pursued, there would, I fancy, be
no need of guessing : a more satisfactory result
would, in all probability, follow. Try it.

Why Dr. Carpenter should substitute guessing
for legitimate enquiry is best known to himself ;

but this method of disposing of a subject which, from its importance demands careful investigation, is what we should not have anticipated of a person so accustomed to scientific research as he is, and who has for many years been an accepted authority on the subjects on which he treats. But he has, in this instance, at least, stepped out of the legitimate tract of exact science, and gone into the path of questionable hypothesis! and "guesses" that the functions of the non-responsive cerebral parts to Faradisation are instruments of purely intellectual operations; from which we infer that thought sets at nought Faradaic-evocation. He proceeds to say :—

"But the negative fact just stated may be considered as a decided confirmation of the conclusion arrived at by the writer twenty-seven years ago, on the basis of comparative Anatomy and Embryology, that the *posterior* lobes of the cerebrum are the instruments, *not* (as maintained by phrenologists) of those passions and propensities which man shares with the lower animals, but of attributes peculiar to man, which we fairly may suppose to consist in such mental operations of a purely intellectual character as do not express themselves in bodily action."—*Mental Physiology*, pp. 714, 15.

According to this view, the posterior and lateral portions of the middle lobes, as well as the whole of the posterior lobes, are centres or organs of the intellect. Now the brains of idiots, in general, are remarkable for the largeness of these parts and the smallness of the anterior lobes; and those persons who are marked for intellectual power, are equally noted for the extraordinary size of the anterior lobes— as a general rule. These facts are so well known, and the concomitance of unusual size with uncommon manifestation of talent are so easily distinguished and capable of verification as not to demand further remark. What gave rise to this peculiar guess of Dr. Carpenter I cannot divine, but the fact seems to me to verify his remarks on the influence of beliefs on the will.

> "'That we easily believe what we wish,' is a proverb which experience shows to be so often true, that science is called on to give the rationale of the fact."—*Ibid*, p. 339.

I shall not attribute to Dr. Carpenter the state of mind which the preceding lines of the paragraph just quoted specify. However, the fact is worth recording :—

> "As soon, however, as *any other motive* than the desire to arrive at the truth enters into the formation of our beliefs, the will comes to have a far more powerful influence."

Notwithstanding the evidence disproving that the posterior cerebral lobes are centres of purely intellectual operations is massive and irrefragable, there is a way in which an important influence is exerted through them, or by means of them, over the intellect according to Phrenology, namely,—by concentrating the attention, by promoting a persistency of effort, self-reliance, self-respect, dignity of bearing, laudable ambition and self-control. Again, had Dr.

Carpenter paid as much attention to the study of
pathognomy as he has to anatomy and physiology,
he would not have fallen into the error of saying
that the posterior lobes are the instruments of attributes that do not express themselves in bodily action ; for pride, vanity, firmness, and fear, express
themselves in a very marked manner, the meaning
of which, infants and animals soon learn to interpret ; and a correspondence between the development of the posterior lobes and the expression of
these emotions is so palpable and so common as to
be a subject of daily observation, that any person
of ordinary intelligence and discrimination, whose
mind is not barred by bias and preconceived
opinions, may soon make himself acquainted with.

Dr. B—— suggests the advisability of my noting
the slender evidence upon which physiologists rest
their beliefs, and found theories on specific cerebral
functions, a suggestion with which I concur ; but
not wishing to unduly enlarge on the subject, I
shall merely note a few facts regarding the theories advanced on the cerebral organs of articulate
language and the bases on which they rest.

Schroeder Van der Kolk, put forth the theory
that "the functional centre of articulate sounds is
the medulla oblongata." Bouilland's theory is that
this organ is situated in the anterior lobes ; and that,
while he admits that speech may exist with one
frontal lobe destroyed, or seriously damaged, articu-

late language becomes impossible. Dr. Dax assigns this function to the left hemisphere without limiting it to any part of it ; whilst Professor Broca limits it to the posterior portion of the third frontal convolution of the left hemisphere.

Now, with Dr. Bateman's exhaustive work "On Aphasia" before me, and an excellent paper "On Impairment of Language," by Dr. W. A. F. Brown, I have abundance of material for exposing the meagre evidence these theories are based on ; but I can only find room for the following extracts. Dr. Bateman says :—

"An impartial sifting of the mass of evidence I have accumulated has led me to the following conclusions :—

"That although something may be said in favour of each of the popular theories of the localisation of speech, still, so many exceptions to each of them have been recorded, that they will none of them bear the test of a disinterested and impartial scrutiny."—*On Aphasia*, p. 178.

Dr. Brown, in his letter to me, July 27th, 1874, says :—

"I think that, in alluding to Broca, and the numerous pathologists who have adopted his views, it is imperative to expose the very slender and unphysiological evidence with which this class of inquirers is satisfied. They conceive that, by detecting structural alteration in a particular portion of convolution in *one hemisphere* when aphasia is present, the connection of this region with the power of articulate speech is established. Now, no phrenologist, the most enthusiastic or the most rash, would have ventured to rest upon such a basis, unless the presumed connection had been otherwise demonstrated either by the correspondence of size with in-

D

tensity of function, or by the existence of disease in similar
convolutions in both hemispheres."

Let us, now, after this apparently lengthy digres-
sion, again turn our attention to a consideration of
Mr. Lewes's charges against Gall's procedure.

In " History of Philosophy," p. 447, Mr. Lewes
represents phrenology as resting on four positions :—(1) " That the grey matter of the convolutions is the organic substance of all psychical
actions. (2) That no other part of the nervous
system has any essential connection with the mind.
(3) That each distinct faculty has its distinct organ.
(4) That each organ is a limited area of grey matter ;" and then goes on to say, that " only one of
his (Gall's) four positions can be accepted as true."
Whatever may be the basis of these positions, it is
not an evidential basis. Mr. Lewes does not only
do Gall an injustice by propounding such fanciful
positions as the basis of phrenology, but he does
himself an injustice also, for no one knows better
than he, that they are at variance with phrenological
doctrine.

Gall asks, " How far does truth bear us out in
saying, that the organs of the soul are situated on
the surface of the brain ?" " Functions of the Brain,"
vol. iii, p. 2. This question taken from the context
might give a pretext to an unfair critic for alleging
that the author advocates the absurd doctrine of
each organ being a limited area of grey matter ;

but such criticism would be unjustifiable, and more especially in the present case, for the author explains his positions with unmistakeable preciseness, beginning by a recital of well-known laws of nature, of which a summary is here given. (1) That the disposition of a large and healthy organ to functional action is greater than a smaller one, other conditions being equal. (2) The nerves are bundles of fibres which are spread out on the parts they are designed to serve—for example, the olfactory nerve is spread out upon the pituitary membrane of the nose; the nerve of taste terminates in minute branches on the tongue; the expansion of the optic nerve forms the retina. (3) The outspread fibres of a thick nerve cover a larger area than a thin one. (4) The size of a nerve may be inferred from its terminal expansion on the area of its ramifications. (5) Nature follows precisely the same law in the brain. The different cerebral parts arise and increase in different places; they form larger or smaller fibrous bundles, which terminate in ramifications. All these ramifications of the different fibrous bundles, when re-united, form the hemispheres of the brain, such as Nature has placed them in their folded state in the cranium. The ramifications of the olfactory nerve forms analogous folds in the alœ of the nose.

A small nervous bundle can form only a small ramification, and consequently but minute folds, and but one or more small convolutions. A considerable nervous bundle, on the contrary, forms an extensive and thick ramification, and, consequently, folds and convolutions of much greater volume.

Thus, then, although all the integral parts of any one cerebral organ, from their origin to their termination, are not situated on the surface of the brain, nevertheless, we can deduce from the size of the fold, or convolutions, positive inferences as to the volume of the whole organ. The longer, deeper, and broader the convolutions are, the more space do they occupy, and the more are they elevated above those that are shorter, narrower, and more superficial ; so that a brain, the integral parts of which have acquired an unequal development, exhibits on its surface depressions, level parts, and eminences.

Mr. Lewes charges phrenologists with having " presented a rude sketch as a perfect science." —*ibid,* p. 430. How does that comport with this ? " I do not conceive that phrenology has reached perfection now, nor do I hope that its application, even when perfect, will always be without error." " Phrenology in connection with Physiology," part i. p. 9, 1826) ; and Dr. Vimont, Mr. Combe, and the Fowlers say the same in substance, and so does every phrenologist of note. More errors of Mr. Lewes might be appointed out, but enough for the present.

Are phrenologists justified, in the present state of

knowledge of the structure and functions of the brain, in assigning to each mental faculty a particular portion of the brain as its special organ? Dr. Gall thought he was, and recent researches justify his conclusion. But whether it was politic to speak of such relationship as an ascertained fact is, perhaps, questionable. The conception was fruitfully suggestive; but had he remained content by propounding it as a highly probable theory, and only spoken positively of indications of special talents, emotions, aptitudes, and tendencies, and omitted for the time being treating of organ and faculty, a deal of adverse criticism would have been avoided, and although phrenology has made rapid strides, it would probably have run a much easier course and a more successful race. Even now, this course is advisable, for, it would appear, that we are on the eve of a revolution in cerebral physiology; and it is barely possible, that the phrenological map may have to be modified, notwithstanding that no substantial reason for such a modification has yet been adduced.

Gall revolutionised the method of dissecting the brain; and his hypothesis that the cerebral hemispheres are not homogeneal in function is now a demonstrated fact. Time has proved it one of those " luminous hypothesis which greatly enriches science." He therefore might have rested satisfied with making it known, and patiently awaited the results of scientific research for its verification.

Medico-psychologists are alive to the importance of ascertaining the functions of the brain. Professor Broca twelve years ago gave an impetus to this inquiry in his researches in aphasia. He concluded that the organ of articulate language is situated in the posterior part of the third frontal convolution of the left cerebral hemisphere. Since then a large number of brains have been examined by anatomists and pathologists, who have called into requisition all the appliances at the command of science to test the accuracy of Broca's observations and inductions.

Phrenologists have always courted investigation, and wished to have their principles put to the crucial test of fact, notwithstanding they may, in their ardour, have manifested impatience, and inferentially overstepped the march of science in cerebral physiology.

Those who speak of phrenology as the physiology of the brain may be right, nevertheless this is not in accord with the general opinion. Whether or not it will ever bear the crucial test of science, lies hid in the future. The knife has failed to reveal what observation and induction have discovered in other departments, and probably will do so again in this case, but we must await the revealments of time. Truth must continue our motto. Striving for victory at any price has dwarfed the growth of many truths, stultified many useful improvements, and strangled others at their birth.

We are greatly indebted to scepticism and conservatism, notwithstanding they often appear to
retard the progress of truth.   Conservatism is to
radicalism what a drag is to a vehicle; and while
the sceptic checks the precipitate movements of the
over-credulous and incautious, the radical goads on
the slow-paced adherent to antiquated things, principles and notions ; and all in turn receive a shock
in the encounter, and thus act and re-act bene-
ficially on each other, and on the ultimate well-
being of the community.   Once let us fully realise
this grand fact, and we shall be inclined to judge
more leniently of those who are not at one with us
in thought.

Mr. Lewes, in his capacity of historian, specifies
the doctrine of phrenology as it was conceived in
1802.   We are told on the authority of Lélut that
at this time Gall had fixed upon twenty-four organs
as representing original faculties ; that he gave to
M. Charles Villiers a plate which represents a skull
with these four-and-twenty organs marked on it ;
that he afterwards gave four of them up, and changed
the localities of the rest, so that scarcely one of them
occupies its original place.   Mr. Lewes then goes
on to say :—

" Phrenologists should give prominence to this fact. They
are bound not to pass it over. For Gall had been twenty
years collecting facts of correspondence between external

configuration and peculiarities of character ; and had controlled these observations by repeated verifications."

He then asks the pertinent question :—

" If Gall could be deceived after twenty years of observation of facts, which, according to his statement, are very easily
observed, because very obvious in their characters, why may he
not have been equally deceived in subsequent observations?"

Now, the revelancy and force of this question
cannot be denied on the assumed facts ; and, having had my attention directed by some reviewers of
" Phrenology and How to Use it," to this portion
of phrenological history, and, not without some
show of reason, I have given the subject considerable attention. I have just finished reading again
the history of the discovery of every organ of Gall
as narrated by himself, and his description of their
localities, and having made other inquiries which
seemed to me necessary for the purpose of testing
the verity of Mr. Lewes's narrative, the conclusion
at which I have arrived, after careful consideration,
is, that the narrative is baseless ; and that the evidence clearly proves, that Mr. Lewes had not taken
the trouble which he was bound to do, as an impartial historian, before citing such statements as
facts for the settlement of such an important issue.
Mr. Lewes's authority is Lélut, and Lélut is said to
have got his information from a letter said to have
been written to Cuvier by M. Charles Villiers. This
letter, then, is the basis of Mr. Lewes's so-called
facts, but he acknowledges not having seen it ; yet

he gives it all the weight and authority of ascertained truth.

Every person who is acquainted with phrenological history knows that the first observations of Gall . were the concomitance between the prominence and depression of the eyes and the memory of words and articulate language ; and they also know, that the signs of these qualities remain the same as they were originally in the phrenological map, so it is with regard to Firmness, Combativeness, Adhesiveness, Love of Young and the Sexes, and in fact the entire list.   That Gall may have seen reason to modify some of his earlier views regarding the outward sign of a faculty is only what might be expected, but there is not a tittle of evidence to prove that he after 1802 altered the localities as he is said to have done.   Had he done so, we have reason for believing from his candour and explicitness that he would not have suppressed the fact.   Spurzheim and Gall were not on the best terms of friendship for some time after 1813, yet Spurzheim never alludes to Gall having altered the seats of the organs, nor does he take the credit of doing so himself, or of assisting in doing so, notwithstanding that if such alteration had been made, he would have known it, for he was associated with Gall, first as pupil and anatomical dissector, and afterwards as joint partner from 1800 to 1813.   Spurzheim was no way backward in asserting his claims, and he takes pains to specify in his own publications the part of the work he performed in the joint productions of Gall and himself on the anatomy of the brain and nervous system, with an atlas of 100 plates ; and, further, had Spurzheim designedly refrained from alluding to this assumed alteration in his publications, it is more than probable that he would have referred to such a notable fact to Combe and others privately. Again, several of Gall's admirers were jealous of Spurzheim eclipsing the glory of Gall, by certain phrenologists attributing to him honours what in justice belonged to his master ; and notably so, Dr. Elliotson, who vigorously supported Gall's claims, and in a style of partisanship that provoked controversy which would doubtlessly have caused the subject under consideration to crop up if it had been true ; but it did not, which is presumptive evidence that no foundation for it existed.

After proceeding thus far, and being still wishful to probe the canker to the bottom, I went to Edinburgh and spent upwards of a week in the Phrenological Museum, examining every thing in it that is calculated to throw light on the subject, and the result confirmed my opinion that the charge is not founded in fact.

Should any person who, having drank at the fountain of Lewes, still be labouring under the effects of the draught, and feel the need of an antidote, it would be well for him to repair to the Edinburgh Phrenological Museum, and to impartially and diligently investigate the evidence stored there.

However, it is only just to remark, that this evidence is somewhat puzzling, inasmuch as the localities of the organs are indicated by different numbers by Spurzheim from what they are by Gall, and not only so, but a different order of numbering was adopted by them at various periods as they progressed in their discoveries. But this is not all : several of the numbers of the organs in the plates, representing the brain and crania in the Atlas published in connection with the joint work of Gall and Spurzheim ("The Anatomy and Physiology of the Nervous System in General") are at variance with the descriptive text. Plates 98, 99, and 100 (the last in the Atlas) represent side, back, and front views of a skull, having all the organs discovered by Gall marked on them in the exact situations as described in the text ; but unfortunately twelve organs in all the preceding plates are incorrectly numbered according to the text.

The two separate numbers that immediately follow the names of the subsequent organs in the table represent the discrepancies just named. The first number points out the locality of the organ to which it is attached, as described in the text, and the second number is engraved on the plates, instead of it ; but both, it will be observed, refer to the same locality, and not to any change that has been made, as some persons might erroneously suppose. Errors of this kind, however, tend to confuse, if not to mislead, casual observers ; but to make them the foun-

dation of such a charge as Mr. Lewes alleges against
Gall, is reckless and culpable.

The Organs.	According to Text.	According to Plates.	The Organs.	According to Text.	According to Plates.
Combativeness ..	4	5	Colour..........	16	20
Destructiveness ...	5	6	Comparison ...	20	24
Acquisitiveness ...	7	9	Benevolence ...	24	26*
Constructiveness	19	8	Veneration	26	27
Self-Esteem	8	12	Firmness.........	27	13
Approbativeness	9	11	Educability† ...	11	21
Number	18	19			

These blunders must be attributed to Spurzheim,
for in p. 14 of his " Anatomy," he claims the credit
of having superintended the drawing and engraving
of the plates, and of writing the description of them.

Mistakes of this kind are, as I have said, puzzling
to the student ; but it would appear that in the present case, they have not as yet done much harm,
and for the simple reason, that they were unknown
previous .to my investigations, or at least before I
caused inquiries to be made respecting Mr. Lewes's
charges.   Notwithstanding, it is barely possible that

* In some plates 14.

† This organ was divided into two by Spurzheim,—the lower portion
called Individuality and the upper portion Eventuality.

this conflict of numbers has, to some extent, given rise to the erroneous charges attributed to Gall of having changed the phrenological topography almost entirely, since 1802. But it surpasses my comprehension how any impartial anatomist could fall into such an error, from an investigation of Gall and Spurzheim's "Anatomy and Physiology of the Nervous System," seeing that the localities of the organs are minutely described in the text in accordance with anatomical nomenclature; much less such experts as Lélut, Pierce, and Lewes, and they were bound in honour by the gravity of the subject to have examined it minutely.

The drawing and marking of the skull said to have been given to M. C. Villiers by Gall, may have been misinterpreted, but his published works should have been honestly appealed to as commentaries for elucidating the obscurity hovering round the conflicting testimony, if conflict there be, between the said skull and its rough markings, and the more elaborate and minute descriptions of Gall.

Engravings 5, 6 and 7 are accurate representations of a calvarium which is in the Edinburgh Phrenological Museum, photographed and engraved to a scale of one-fourth the natural size. This calvarium is valuable as showing the state of phrenological topography in 1806 or 1807, by which, it will be seen, that the localities of the organs are the same now as they were at that time, although the numbers indi-

cating their situations are very different. This being the case, and supposing for argument's sake, that
Gall did abandon four organs, as representing original faculties, and so altered the remaining twenty,

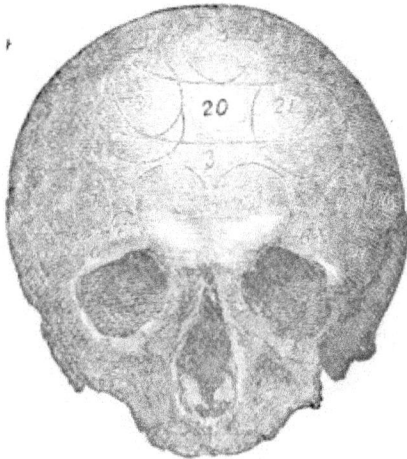

No. 5.

that scarcely one remains in its original place,—I
say, supposing this were true, these acts must have
been performed between 1802 and 1806 or 1807,
which, to say the least of it, is very improbable.
Gall was not a likely person to make such rapid
changes; to feel himself justified in so short a
period in pulling down the fabric that had taken
him twenty years to build up, at an enormous ex-
penditure of brain force and money. No, no, he
was not such a careless architect, nor was he so
reckless of his prestige as to build it on a sand-drift
of dubious theories, instead of on the rock of induc-
tive certitude.

The subjoined notice of the calvarium just alluded to, is extracted from Mr. G. Combe's "Notes on the United states of America during a Phrenological Visit," in 1838–40, vol. i. pp. 304, 305 :—

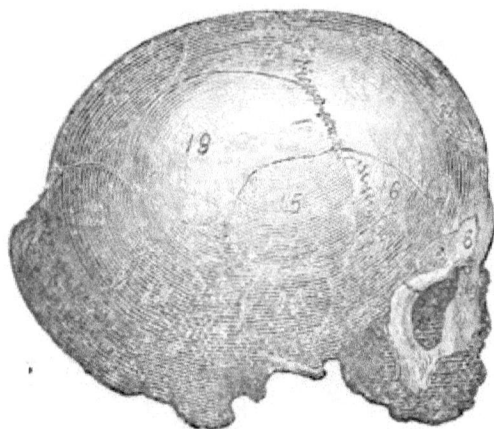

No. 6.

"*January 4th, 1839.*

"Mr. Nicholas Biddle, Manager of the United States' Bank, called and informed me that he had attended a course of lectures by Dr. Gall, at Carlsruhe, in Germany, in 1806 or 1807, and he presented to me a skull which Dr. Spurzheim had marked for him shewing the situation of the organs as then discovered, and which had remained in his possession ever since. This relic possesses historical value. It has been often asserted that Gall *invented* his physiology of the brain, and did not discover it.

"When I was in Germany in 1837, I saw a collection of books describing the science at different stages of its progress, and also skulls marked at different times ; all proving that the organs were discovered in succession as narrated by Drs. Gall and Spurzheim. This skull which records the state of the science in 1806 or 7, presents blank spaces where the or-

gans of Hope, Conscientiousness, Individuality, Concentrativeness, Form, Size, and Weight, are now marked, these
having at the time been unascertained. Farther, the local
situations, and also the functions of the organs then marked
by Dr. Gall as ascertained, continue unchanged in the marked
skulls of the present day."

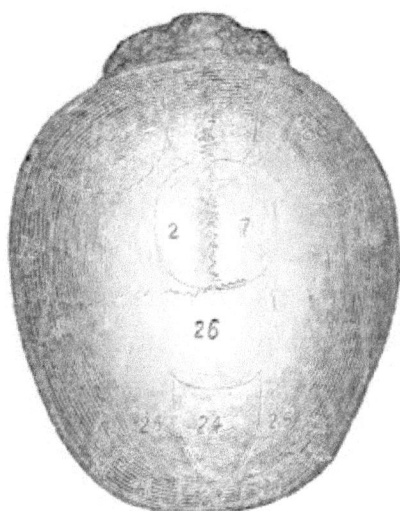

No. 7.

Another objection brought against phrenology is,
that it leads to fatalism, and its doctrines are anta-
gonistic to Christianity. This is an old objection,
and although it is successfully combated by Gall,
Spurzheim, Combe, and others in their works, it still
endures the ravages of time, and often appears on
the stage with all the vigour and boldness of youth.
However, it is more formidable in tenacity of life
than potent in argument, and is the least tenable of
all the objections. If it has any force. it applies
equally to every system of mental science, for they

all have a physical basis, and the body and mind are mutually dependent on each other.

"Phrenologists teach," say some persons, "that man must necessarily act in a certain manner in accordance with his cerebral development." I reply, that phrenologists do no such thing. What they say is this, It has been found by multiplied observations, made on almost all classes of people, under very various conditions, by intelligent and trustworthy inquirers, capable of forming correct judgments, that there is a general correspondence between particular forms of head and certain specified mental tendencies and talents; and they believe themselves capable of inferring the natural inclinations of persons from the forms of their heads with a tolerable degree of exactness, and, likewise, what any person *may* become under suitable conditions; but,—and mark this well,—*not what he will become.*

Those who desire to see this objection completely disposed of are referred to the above authors; and my brochure on the "Characteristics of the Rev. C. H. Spurgeon."

E

CHAPTER II

EXPOSITION OF THE WILL.

THE fact of no local habitation being assigned to the Will in the phrenological map, is advanced as an argument against the completeness of the system. On this I would remark, that Gall did not consider the Will a single faculty having a special organ.

"Ce n'est point l'impulsion resultant de l'activité d'un seul organe, ou comme disent les auteurs, le sentiment du desir, qui constitue la volonté. Afin que l'homme ne se borne pas à desirer, pour qu'il veuille, il faut le concours de l'action de plusieurs facultés intellectuelles supérieures ; il faut que les motifs soient pesés, comparés, et juges. La decision resultant de cette operation s'appelle la *Volonté*."— *Fonctions de Cerveau*, par F. F. Gall, tom vi. p. 428. (1825.)

[TRANSLATION.]

"Will is not an impulse resulting from the activity of a single organ, or, according to certain authors, the feeling of desire. In order that a man may not limit himself to wishing in order that he may will, the concurrent activity of several of the higher intellectual faculties is necessary ; motives must be weighed, compared, and judged. The decision resulting from this operation is called *Will*."

The operations of the Will, and its influence, mentally and physically, on the workings of the mind, and its outward manifestations; the influence of motives on the Will, as well as the various emotions and intellectual proclivities and their causes, present a varied and an extensive field of inquiry far exceeding the limits of this work. Seeing, then, that we cannot traverse this field, let us take a view of it from the most convenient standpoint we can command, and endeavour to ascertain whether or not Gall's conclusion be correct.

No purely voluntary action can take place without the Will, hence the terms Volition and Will are used synonymously. The term Will is also used as expressive of a resolve, a decision to do something. or perform some act on a future occasion. It is likewise made use of in a compound sense, designative of traits of character, such as disobedience,—as a self-willed child; determined persistence,—as a wilful, headstrong man; definitiveness of aim and oneness of purpose,—as a strong-willed person; deficiency in constancy of pursuit, or unsteadiness of will, —as a person of a versatile mind, who shows a vacillating, changeable disposition, and sticks to nothing long, like a butterfly flitting from flower to flower, and tasting a little of the sweetness of each, but exhausting none. Such a person is said to be weak-willed; and those persons who manifest indecision of character, and are simply led by others, are represented as possessing no will of their own.

These varied uses of the term will tend to bewilder the minds of persons who have not studied the nature of Will proper ; consequently, whilst every one speaks of the Will, comparatively few have a right conception of it.

1. What, then, is the Will, and what are its powers?

2. Is it a faculty of the mind, possessing a special cerebral centre, and being subject to the same laws as other centres ?

3. Is it an independent, self-controlling property, or entity possessing the power of dominating over the mind, and of acting with perfect freedom? or is its freedom limited by circumstances over which it can exert no control ?

4. Is it a distinct faculty? or is the term Will merely expressive of voluntary action and control, as resulting from certain antecedent mental operations that are determined by sense-impressions, and the perceptions, and ideas, and desires that are produced in the mind by these impressions? in other words, do the mental phenomena that we attribute to the Will result from our whole mental life ?

The subsequent remarks and examples of the operations of the Will, crude and ill-digested as they are, will assist the student in solving these questions, and may stimulate him to thoughtful inquiry, and direct his attention to the standard authorities in psychology for fuller information. I simply aim at a popular exposition from a phrenological point of view, believing this method will be the most useful to my class of readers.

Let us inquire what are some of the primitive manifestations of the Will in the brute, of which the dog will serve as a sufficient example; and note a few of the earlier operations of the Will in man as observed in childhood, and its development towards maturity.

A dog, feeling the pinch of hunger, seizes hold of and devours a pullet, for which he gets severely whipped. Hunger, on a future occasion, gnaws at his stomach, and observing some defenceless pullets, he longs for one, but remembering his punishment for his former misdeed, he restrains himself. What is the cause of the change in the animal's conduct? It is clearly the fear of punishment,—supposing his sense of hunger to be as acute in the latter case as it was in the former. Here we see the effects of two opposing forces: hunger and fear contend for mastery, but the latter comes off victor, and determines the dog's choice and his action,—in other words, determines his Will. At another time his dinner is put down to him when he is resting. The flavour of the meal arousing his consciousness, he rises and looks at it for a while, then resting on his haunches, takes another look, and finally lies down again without touching the food. Here, again, we observe two forces have been contending: a desire to eat and a desire to rest, in which the latter gained the ascendancy, and the Will acted accordingly.

I owned a dog of the Newfoundland and re-
triever breed, and I was much interested in observing the operations of his mind. He was very fond
of bathing, and often spent a whole day in this indulgence. Through knowing his habits, and by studying his expression, I could perceive by its indications
when he proposed going to bathe ; and I often succeeded in thwarting his intention, and at other times
in drawing him from his favourite pursuit. " Albert,"
I would say sometimes, " I see you are going to
bathe ; now, the dinner will be ready at one o'clock,
and if you do not return at that time, you will not
get any." He would look me in the face, and giving unmistakeable signs that I was understood,
would steal slyly off ; but notwithstanding, I do not
recollect of having talked to him to this effect in
vain. It was frequently remarked by myself and
other members of the family when the cloth was
being laid, that " Albert would turn up just now,"
for he usually, under the circumstances, came at the
appointed time. I fancy I see him now standing
with his forepaws on the window sill, at the outside
looking into the dining-room, notifying his return,
and asking for admission. The thought of his dinner had determined his Will from either not going
to bathe, or not indulging in this luxury so long
as he was wont to do.

An analysis of these examples of the operations
of the dog's Will, shows that the initiative in the first
case was the impulse of hunger ; that in the second

case, hunger was restrained by fear, and this was
excited by experience or the recollection of punishment; then judgment decided not to incur the like
again; and the Will carried out its decision. In the
third example, the opposing motives were so poised
for a time as to cause indecision, until the desire
for rest turned the scale, and the Will acting in accordance therewith, caused the animal to lie down.
The last case shows a higher manifestation of mind,
a more complex operation of the knowing faculties,
of memory, experience, and trust : the evidence
presented to the judgment being greater and more
complicated for the guidance of the Will, and all the
examples show that the dog's Will was not a self-
determining power, but the resultant of feeling,
memory, and experience, or that these were the re-
mote causes, and judgment the proximate cause, of
the resultant actions,—the Will being simply the
executor of the judgment, the liberator of the nerve
force which set the voluntary organs in motion for
the accomplishment of a definite purpose; much in
the same way as an engine driver opens the valves
of the engine, and liberates the steam pressing
against them, so as to allow it to flow into the
cylinder, and sets the engine in motion for a specific
end.

If the dog were endowed with a moral nature, so
as to feel a lively sense of responsibility for his
actions, and that doing right were essential to his
personal happiness as well as that of his species, and

wrong doing entailed misery, other motives arising
out of this sense would be laid before the judgment
• in deciding upon courses of action involving the
moral sense of right and wrong, and greater delibera-
tion would be necessary.   But if the animal were not
fitted with a higher endowment of intellect to acquire
information, and to judge of evidence commensurate
with the moral sense, he would be in a worse position than he now is.   His Will would be less self-determining and controlling ; inasmuch as the desire to
do right, and the fear of doing wrong, would trammel
and weaken the judgment, and consequently the
power of the Will.

By attentively studying the operations of the
human Will, from the first dawning of intellect to
maturity and old age, we find that it is governed by
similar laws as those of the brutes', with this important difference, that man is endowed with a moral
nature, and an intellect in keeping therewith, consequently his Will is influenced by a more numerous
and higher range of motives.

On presenting to a child for the first time a piece
of alum and a piece of sugar, he instinctively seizes
hold of both, and puts them to his mouth, when he
finds the one agreeable and the other disagreeable :
this makes an impression on his mind, and by repeatedly presenting these substances to him he
learns to discriminate the difference between them,
and ultimately refuses the alum and accepts the

sugar, which indicates decision of judgment as the basis of his Will; whereas, in his initial trial, his action was the result of instinctive impulse, arising from curiosity, or other simple or complex impressions.

Should he take ill, and a doctor be called in, he afterwards remembers his pains, and that the doctor looked at his tongue, and felt his arm (pulse). He likewise recollects that he took physic, or something nauseous; and for a while after the medical attendant has discontinued his visits, the child associates his ailment, the examination of his tongue and pulse, and his taking medicine, with the doctor, and not being capable of distinguishing other gentlemen from him, he becomes alarmed at their visits, and refuses to put out his tongue and let his arm be felt by them, and to take anything out of a spoon or other utensil by which the physic was given to him, until he learns to discriminate between the doctor and other persons, and the medicine and the things pleasant and unpleasant to his taste, and so is able to dissociate in his mind the connection between his ailment and the doctor, and every thing pertaining to him. In this case, the boy's Will is influenced by the stronger motives and experiences similarly to the dog's.

In the matured and educated mind of larger experience we observe other emotions coming into play, and presenting a different order of motives for the decisions of the judgment, by which the Will is prompted to action,—such for example as emotions

of justice, truth, charity, sympathy, tenderness, reverence, dignity, ambition, love, and attachment, a
desire of knowledge, love of refinement, utility, the
welfare of mankind, self-interest, self-abnegation,
duty, and obligation to God and man, etc.   To
analyse the effects of these various elements on the
mind in giving birth to motives, and showing where-
in the stronger prevail with the judgment, and
prompt the Will to act in a certain way in preference
to another, might be interesting, but this is beyond
the scope of the present work.   A few examples,
however, will show that, in these more complex
operations of the mind, the Will is subject to similar
laws as it is in the case of the dog and that of the
child.

The Will is said to be the highest force of the
mind, notwithstanding it has not unlimited control
over the body, as is shown in the case of paralytics ;
neither is it all powerful and controlling over the
mind.   We cannot Will to sleep or wake, to think
or not to think, to remember or not to remember,
to love or to hate, to feel gay or sorrowful, to express or not to express by word or gesture the
various emotions that pervade the mind, etc., when
we like ; nor can the Will originate an idea or dismiss one at pleasure when it has taken possession
of the mind.   We may, by change of topic, com-
pany, and scenery, gradually get rid of a trouble-
some idea, but to dismiss it at once by an effort of

the Will is frequently beyond our power ; and the Will is equally powerless for determining the materials of thought, nor can it execute movements for a special aim before the habit of such movements be acquired, and when the habit is acquired, the Will cannot cause it to be forgotten until the judgment has decided to give it up, and this decision can only be arrived at by presenting to the judgment a stronger motive for desisting than for continuing the practice of such movements ; and then, it takes time, less or more, to forget the habit proportionate to the persistency of the muscles to act in accordance with it ; hence the absolute necessity of right training, education, and proper associations.

Supposing an intelligent, thoughtful, and just man to be considering the best way to spend his holidays, and to have several places in view, each possessing specialities for his choice,—his aim being to get as much healthful recreation as he can, compatible with limited means and delicate health, and the interests of his wife and children, whom he purposes to take with him. It is obvious that his purse, his health, the time of the year, and the comfort, travelling capacity, and enjoyment of his family would present special features for consideration, and the strongest would be likely to determine his choice ; and it is equally obvious that to arrive at a right judgment he should be unbiased, unselfish, and have correct information as to the routes, accommodation, objects of interest, the

means of getting to them, the salubrity of the atmosphere, and the probable benefit and pleasure each
• of the districts would be likely to afford, and his
decision would necessarily be in accordance with
the balance of evidence.   Such would be the result
in the case of a person having decision of character, and if he had firmness of purpose, the verdict
would be carried out, circumstances permitting ; but
if he were an undecided person, he would have
difficulty in coming to a decision proportionately to
his deficiency of this quality of mind, and, when he
did decide, should he be inconstant, no reliance
could be placed on his carrying out his resolves.
A person of this kind would be properly described
as wanting in self-control.   But what of his Will ?
Would his indecision and vacillation be attributable
to unsteadiness of the Will ? or to weakness of judgment and infirmness of purpose ? or to the constitution of his mind in general ?   To the latter, I
think.   For the judgment might be trammelled by
Apprehensiveness, deficient Firmness and power of
concentrating his attention, and Love of Change, or
all of these acting together with an equal degree of
power.   The same qualities would tend to bring
about a change of purpose after a decision had
been come to, as one or more of them happened
to be overcome by the other, whilst the Will waited
in readiness to carry out the decision, like a rail-
way engine-driver waiting in readiness to move the
engine as soon as he receives the guard's signal.

The Will seems to be no more self-determining and capable of over-riding the judgment than the hangman is to over-rule the judgment of the court in a criminal trial. But, as the office of the latter is to carry out the law, so is it the office of the Will to execute the decisions of the judgment. Yet the parallel does not hold throughout; for, whilst the hangman cannot alter the evidence on which the condemned is convicted, we can by efforts of the Will alter the circumstances which give rise to the motives that influence our judgments. But here, again, motives take the precedence, for we must have reason for altering the circumstances, so as to induce such volitional control, which presupposes motives as the basis of our reasons. For example, a " fast young man," brought to consider the error of his ways, by affliction or some other sudden calamity, resolves upon a thorough reformation of conduct, and he puts forth all his might to retrieve his character. But he finds the giving up of old habits, and the breaking off of long-established con-nections, hard work. Yet, being thoroughly im-pressed with the necessity of saving himself, and of doing it at once, he summons all his powers to do battle with the enemy. He no longer trims with him : conquer or die is his motto. Then old companions are given up, and new ones sought more in keeping with his altered state of mind. The library and the lecture room are substituted for the drinking saloon : Works of fiction and romance are replaced by treatises on science and morals : Literature and religion, sociology, self-discipline, self-reformation, and the good of society, form subjects for thought and topics of conversation ; and by concentrated and persistent effort, he ultimately extricates himself.

This person has altered the circumstances which influenced his motives of action, his surroundings, trains of thought, and general habits ; but the starting point was the stronger motive,— the offspring of the sudden awakening from the slumber of passional thraldom.

The latest exposition of the Will is from the pen of Dr. Carpenter (" Principles of Mental Physiology"). He speaks of it as being free and the " self-determining power," a " something essentially different from the general resultant of the automatic activity of the mind" (p. 392), by the exertion of which each individual becomes the director of his own conduct ; "and so far," says he, " the arbiter of his own destinies in virtue of its domination over the *automatic* operations of the mind, as over the automatic movements of the body ; the real self-formation of the Ego, commencing with his consciousness of the ability to determine *his own* course of thought and action." But this self-determining, self-directing power is not, as it would appear, according to the author, an innate power, but it has to be acquired. He remarks, " until this self-directing power has been acquired, the character *is* the resultant of the individual's original constitution, and of the circumstances in which he may have been placed, his character is formed *for* him, rather than *by* him, and such a being, one of those heathen outcasts of whom all our great towns are unhappily but too productive, can surely be no more morally responsible for his actions than the lunatic who has lost whatever self-control he once possessed."—Pp. 9, 10.

That man's character and beliefs are, to a certain extent, moulded by circumstances, and that he has the power of altering some of the circumstances, and is to this extent the arbiter of his destinies, are facts generally admitted, and so far Dr. Carpenter's theory is in accord with the general belief. But the doctrine that persons who pander to their appetites, and lead immoral lives, are no more responsible for their actions than lunatics, is at variance with the opinions and laws of all civilized peoples and nations.

Are there any of those heathen outcasts of whom all our great towns are but too productive, who are sane, that have not self-determining power? Persons that are deficient in decision of character and firmness of purpose, who are vacillating, changeable mortals, and are tossed to and fro with every wind of doctrine,—creatures of impulse and excitement of the moment,—are often spoken of as not having a Will of their own. But this mode of speech is not to be interpreted as literally and scientifically true. All that is meant by it, or ought to be understood by it, is instability and deficiency of self-discipline and control. Again, can it be legitimately said of those who lead immoral lives, that are frequenters of the ale-bench, the gambling table, and indulgers in low, brutalizing pastimes, that they are no more responsible for their actions than lunatics? Certainly not. If we observe their actions, and listen to their debates, the laying down of their plans, and the arrangement of their over-reaching gambling plots, we shall see striking indications of sagacity, dominant Will, and self-determining power, besides remarkable tenacity of purpose and unswerving determination, which, if applied to moral elevation, and literary and social distinction, would be eminently praiseworthy. It cannot, therefore, be said that they are devoid of Will; but we may justly say they are deficient in moral power, and, consequently, lack the higher motives and inclinations to distinguish themselves by ennobling characteristics.

Before proceeding farther, it may be as well to state, that Dr. Carpenter teaches that, "The actions of our minds, *in so* far as they are carried on without *any interference* from our Will, may be considered as Functions of the Brain,"—that is to say, they are automatic actions; and such actions, by-the-by, are much more numerous, according to Dr. Carpenter, than we have hitherto been taught to believe them to be. On the other hand he says :—

"In the control which the Will can exert over the direction of the thought, and over the *motive force* exerted by the feelings, we have the evidence of a new and independent power, which may either oppose or concur with the automatic tendencies, and which, according as it is habitually exerted, tends to render the Ego *a free agent.*"—P. 27.

This passage, translated into phrenological language, simply means, that man is endowed with a threefold nature, or with a mental organism by which he manifests this triple condition,—namely, animal propensities, moral sentiments, and intellectual faculties, and that, when the animal propensities largely predominate in any person, we find that he is strongly inclined to appetital impulse and animal gratification in general ; that he feels self-discipline very difficult : a work requiring an ever watchful eye on his natural desires, his words, and actions, and the absolute necessity of having his duty, as a responsible being to God and man, constantly before his mind, in order to stimulate him to determinate and persistent efforts to keep his appetites in check, so that he may by divine aid overcome. Truly may such a person, when fully alive to his state, exclaim, " The spirit is willing, but the flesh is weak."

No person who is highly endowed by nature can conceive the reality of such people's state of mind, —the intensity of their conflicts with their propensities, or have more than a faint conception of the

F

power of temptation, and their susceptibility to it in their case; for as the magnet turns to the north so do their propensities naturally incline to sensual indulgence. This class are the wayside and rocky-ground hearers.

When the moral sentiments dominate, a moral life is easy. Yes, very easy indeed, as compared with the case of the last-named. Persons so endowed represent the good-ground hearers. There is depth of moral-soil, and the seeds of righteousness take root, and "spring up and bear fruit an hundred fold." To those persons in whom the intellectual faculties are dominant, knowledge usually forms the most agreeable mental pabulum.

The great desideratum is to gain moral ascendancy under the guidance of an enlightened understanding; then the self-determining power will develop into self-control; and what is good and ennobling is the natural outcome, as it were, of such minds. This is the class to whom Dr. Carpenter, I apprehend, attributes the over-ruling power of the Will,—whose volitional actions largely predominate over the automatic, thereby rendering the Ego free; and it is vice-versa in the case of the rocky-ground hearers. Yet I cannot allow the justifiability of denominating this latter class as having no Will; although it may in truth be said, that the Ego in such persons is fearfully trammelled in moral action.

We are told that all the actions of early childhood are automatic :—

"The more carefully the actions of early childhood are
observed, the more obvious does it become that they are
solely prompted by ideas and feelings which automatically
succeed one another in uncontrolled accordance with the laws
of suggestion."—P. 264.

According to this view, a child has no Will.
This is doubtless true of new-born infants; but at
what period a child acquires a Will, the doctor does
not state, nor by what means it may be acquired.
However, we may safely say, that Will appears
simultaneously with Reason, and as the power of
reason becomes stronger by the growth of knowledge,
the Will gains strength in like ratio. But does self-
control keep pace with this increase of Will-power?
My observations of the actions and self-government
of children incline me to the negative on this question, and reason corroborates observation,—for proper self-control requires a tolerable balance of the
intellectual faculties, moral sentiments and propensities, naturally, and that they be well-trained
by experience and culture. So Will and Self-control
are distinct, although we cannot possess the latter
without the former.

Dr. Carpenter says :—

"Even in the adult, the predominance of the *automatic*
activity of the mind over that which is regulated by the *Will*,
is often seen as a result of a want of balance between the
two; arising either from the excessive *force* of the former, or

from excessive *weakness* of the latter. We have it in the loose rambling talk of persons who have never schooled themselves to the maintenance of a coherent train of thought, but are perpetually ' flying off at a tangent,' sometimes at a mere sensorial suggestion (conveyed by the sound or the visual conception of a word), sometimes of an ideational association of a most irrelevant kind."—Pp. 265, 6.

Coleridge and Mozart are cited as examples of this order of mind. A rather lengthy account of the lives of these famed men is given, with interesting anecdotal illustrations of the workings of their minds. Coleridge is described as being woefully deficient in self-determining power of Will, in fact, as being little more than an Automaton, —a waking dreamer, a sort of unconscious elaborator of thought,—who, for brilliancy and power has rarely been exceeded, and "there was perhaps not one of the last generation who has left so strong an impress of himself on the subsequent course of thought of reflective minds engaged in the highest subjects of human contemplation," notwithstanding, "it used to be said of him that whenever either natural obligation or voluntary undertaking made it his duty to do anything, the fact seemed a sufficient reason for his not doing it." These characteristics indicate a low state of morals, and imperfect appreciation of obligation and duty.

In a mind so constituted, the motives arising out of moral feeling would be weakly represented, and, consequently, the Will would not be guided by such considerations, and it would be left at the mercy of stronger and more persistent impulses, causing Coleridge's demeanour to be "expressive of weakness under the possibility of strength," as Carlyle aptly describes it. Carlyle further says :— "Nothing could be more copious than his talk ; besides it was talk not flowing any whither like a river, but spreading every whither in inextricable currents and regurgitations like a lake or sea ; terribly deficient in definite goal or aim."

Now, what of this erratic genius, whose brilliancy shone like a cascade *pregnant with sunbeams*, and whose power left an impress on the sphere of thought as indelible as the upheavings of Etna on terra firma. He certainly was a notable example of want of self-determining power ; but surely all this brilliancy and this power were not automatic resultants ; spontaneous and uncontrollable discharges of nerve-force from the hemispheric cortical cells ? Whence came his extensive and varied acquisitions ? the ingatherings of the harvests of literature and art of past generations and contemporary tillers of the soil of thought stored in his mind ? All this volubility of which Carlyle speaks,—excellent talk, very,—was surely not a mere bubbling from the springs of intuition. There must have been toil in the gathering and garnering of the harvests,—aye, laborious toil, either forced or voluntary, and if voluntary, self-denial, self-determination, and strong efforts of Will.

It requires a much stronger Will, a much greater power of self-discipline, to govern a versatile intellect than one of less endowments. The man with one or two strong predominant powers runs in a groove like an automaton; but men whose powers of acquisition are so numerous, strong, and active as Coleridge's, each craving with an insatiable thirst for mental pabulum, require a comparatively superhuman power of self-discipline to keep them in check, so as to be forced, as it were, to indulge in their specialities. Men's power of self-control should be measured according to the strength and activity of their varied inclinations, the plenitude and power of their motive-springs, and the intensity of their cravings.

Mozart is instanced as a typical example of the spontaneous or automatic producer of musical conceptions. He, " like Coleridge, was a man whose Will was weak in proportion to the automatic activity of his mind, hence his life becomes a most interesting study to the psychologist." Yet Mozart is represented (and the account he gives of himself confirms the accuracy of the representation) as having possessed an excellent memory, and complete mastery over his thoughts, likewise a remarkable capacity for methodising and defining them, so that as a sportsman would empty his well-filled bag and divide into separate lots the varieties of game : he could take his thoughts out of his *mental pouch* and select and arrange them, and could "write and talk during the process." "When he was in the vein for composition it was difficult to tear him from his desk." Mozart, then, manifested very diverse traits from Coleridge. He evidently had more than an ordinary share of self-determining power, and why he should be presented to us as a man of weak Will is best known to Dr. Carpenter, but he gives us an inkling of his reasons for doing so in the following passage :—

"If the self-discipline, which Mozart so admirably exercised in the culture of his musical gifts, had been carried into his moral nature, so as to restrain the impulses of his ardent temperament within due bounds, and to prevent him from consuming the energy of his frail body in the pursuit of exhausting pleasures, the world might have profited by a still higher development of his genius, and a still larger bequest of treasures of pure and elevated enjoyment."—P. 275,6.

What the author means by moral nature here is evidently the religious.

Dr. Carpenter speaks of strong Wills, of weak and unsteady Wills, and of persons possessing no Will, and, as we have seen, that the Will—the self-determining power,—has to be acquired. Now all the numerous examples which he gives of the working of these varieties, show that the Will in each case acted with the aim of attaining the greatest apparent good ; that is, what appeared to the mind of each most calculated to yield the greatest amount of present or future good. I say apparent good, for a person may deny himself of present pleasure with

the view of realising greater good in the future, or he
may prefer the pleasure of the moment, to the practice of self-denial and to waiting for that which his
judgment dictates would bring about more real and
permanent good at some future period. This is the
usual course of the epicure and the drunkard, and
those who pander to their appetites. Again, a person may act against both his present and future
interests; and this may arise either through ignor-
ance, want of forethought, or circumspection, or
from other causes.

Coleridge evidently did not conduct himself properly, nor did he accomplish what might have been
expected of him. He had extraordinary talents,
but they were not properly directed and controlled.
He was deficient in persistent effort, and his conduct
indicated defective morals; and that is just such a
character as would have been inferred of him by a
competent, practical phrenologist, from the form of
his head.

I possess two casts of his head, one taken in 1828,
and the other in 1834. The size of his head according to these casts is large, being 157 cubic
inches by water measurement; and the basal region
is by far the largest. The length of it is 8·1 inches,
measured from the super orbital ridge to the occipital spinous process, whereas the coronal region is
only 6·7 inches long between the points of ossification of the frontal bone, and about ¾ of an inch
above the apex of the occipital bone, and the head

rounds off archwise considerably above these points : and the part where phrenologists locate Concentrativeness or Continuativeness is comparatively small. The circumference of the head at the base is 23¾ inches, and of the superior part 21¾ inches. These measurements indicate a vigorous but very unequal balanced mental organism ; but I am digressing.

It would appear, as far as I can make out, that the cause of the distinguishing characteristic of those different varieties of Will, according to Dr. Carpenter, is the comparative strength or weakness of the moral feelings ; and, that Non-Willed persons either possess no moral feeling, or such a very weak moral nature, and are so disproportionate to the strength of their animal and appetital impulses, as to render them, owing to this great want of mental balance, slaves of the latter. In other words, their actions are physically and ideationally automatic, not voluntary ; the former being reflected by the spinal axis, and the latter by the cortical substance of the cerebral hemispheric ganglion.

The following case shows the influence of the emotions on the Will :—

A young man struck with Cupid's dart, and having been accepted by the object of his affection, paid his addresses to her for some time, during which he observed many traits that led him to perceive she would not make him a suitable companion for life, and he pictured to himself a home the reverse of a happy one should he marry her : that disorder, improvidence, debt, and a number of other things would probably mar his happiness. So he tried to break the connection several times, but only to find his own weakness. After repeated intervals of separation, he was always drawn back again by a power he could not understand or control. He saw the whirlpool of misery and pulled hard to steer clear, yet found himself drifting into it, and, at last, was hopelessly engulphed, to the great grief of his parents and friends. Never, perhaps, did a poor fellow find stronger reasons for regretting his course. All his forebodings were more than realized,—in fact, multiplied without end, and intensified beyond human endurance. He suffered and endured until a separation was deemed the only plan of saving himself from committing some desperate act. Separation was effected, and he went to sea for a long trip, after making ample provision for his wife. He now thought the end of his trials had come at last. Vain, however, were his expectations. He returned safely to the home of his parents,—a perfect paradise as compared to the hovel he had left. Another voyage was performed ; and he returned, but not to the comfortable home provided for him by his parents, but to the bed of his tormentor, and here he found additional painful reasons to regret this course. Twice afterwards he flew from his wretched abode for awhile, but only to go back, *and drink himself to fulness of misery*, until death cut the gordian-knot, and placed his tormentor beyond the region of hatching degradation and everything calculated to render life a burden. Let us bring into the foreground the puerile efforts of this man's Will to prevent him from going into the vortex of ruin, which he evidently saw before him, and from extricating himself from it, when an opportunity presented itself.

I know this man well, and was privy to his courtship, marriage, and trials. He is highly respected by all who know him, and is a splendid workman, and by no means in general what may be termed a weak-willed person; yet, in the hands of Cupid, he was powerless. This case is irreconcilable with the doctrine of the self-directing power of Will; for it was not an error of judgment which impelled this person on to the goal of misery. He foresaw the evil, and his friends clearly pointed it out to him likewise. What, then, caused him to drink, so to speak, the poisonous draught? It was the influence of the emotion of love which doubtless suggested to his mind cause for hoping that things might turn out better than he feared they would do, and as his judgment indicated. Then his natural kindness,—for he is acknowledged to be most kindly disposed,—probably caused him to shrink from the idea of so wounding the feelings of the loved one, as he would necessarily do if he abandoned her. Again, he is a just man, and love of fair play would suggest the idea of the unfairness of abandoning her after having won her affections, and leaving her in distress or agonising despair. These emotions of hope, sympathy, and justice, acting in combination with the impulses of Cupid, would tend to produce a balance of evidence in favour of consummating the engagement, and would influence the Will to execute it.

I am also intimately acquainted with a man, who, at the age of six years, gave marked indications of self-denial, self-determination, and power of Will.

It was customary amongst children where he was brought up, to call on their uncles and aunts, and other relations, in the neighbourhood, at Christmas, to get Yule cakes, and for New Year's gifts on New Year's day, and for eggs at Easter; but Master A——, from his sixth birthday, or it may be earlier, invariably refused to accompany his brother and cousins on these occasions, not because he did not desire these gifts, but because going for them seemed to him humiliating and indicative of want of self-respect, except in pursuance of a special invitation. Therefore, rather than demean himself by such acts, he elected to do without the presents.

Ever since that tender age up to the present time, when his crown is whitened by the snows of many winters, he has shown the same characteristic traits, remarkable self-denial and unswerving adherence to principle, for which he has had to suffer very much. He is very abstemious in his habits, and cannot be tempted to eat or drink anything which he thinks unsuitable for his health, however strongly the palate may crave for indulgence. He has passed by the choicest dishes of well-spread tables, which were very tempting to his palate, and only partaken of the plainest fare; and such is his directness of purpose, that I verily believe if a whole street were on fire in a city it would hardly be sufficient to divert his attention from his aim so as to make him visit it, except he could render any assistance; then his own business, for the time, might go to the dogs.

If Will is a self-determining power, a property of the mind superior to the strongest motives, this person surely possesses it; since he manifested it even before he was six years old; yet, if he had not known something of dietetics, and noted the effects of particular kinds of food on his health, he would doubtless have satisfied his palate, when tempted to do so, at the expense of his stomach; hence his conduct is marked by self-control; but this quality, so far as concerns hygiene, is guided by experience. His juvenile conduct regarding presents indicates active Self-Esteem and Firmness as the over-ruling motives.

The cause of the difference between strong-willed and weak-willed persons appears to arise from in-equality in the strength of predominant faculties, —a greater sensitiveness to certain impressions than to others, and to external and internal stimuli,—

also, the quantity and quality of a person's experience, education, knowledge, and surroundings.

Some children (and " men are only children of a larger growth" and experience) are intractable, disobedient and self-willed—that is to say, they are guided by unbending determination, independence and self-interest, rather than the advice and experience and interest of their parents, guardians and tutors, while others are obedient and differential, and manifest a more amiable and lovable disposition. These differences, according to the doctrines of phrenology, are indicated by their external forms, especially that of their heads.

SUMMARY.

1st. Self-introspection and analysis of the operations of our own mind show that Willing is a different function from Feeling and Thinking, yet to think is a voluntary act, as well as a physical movement, that is exerted for a specific aim.

When we will to think, or to move, we do not think of the special organs and their connection, or of the modes of operation by which our thoughts and movements shall be performed, but we merely Will the event, and direct our attention to its production. Being conscious of possessing the power of doing, or of trying to do, what we wish to accomplish, we simply will to do it,—we turn the steam on, as it were, and the bodily machine moves as desired ; but should either the steam (nerve-force), or the machine be defective, the result will be defective too.

2nd. We cannot alter our nature ; and our actions are necessarily limited by natural laws : nor can we help feeling, nor having such desires and inclinations as that feeling produces. The kind and degree of intensity of our feelings depend on the nature of our organic constitution, and on internal and external stimuli.

3rd. The constitutions of people are not alike, but differ materially ; consequently, the feelings, inclinations, and intellectual aptitudes, the power of the Will and of self-determination, of no two persons are alike.

4th. The Will is not a single faculty, having a distinct cerebral centre for its organ, but is a mode of operation of the mind, the actions of which are determined by motives.

5th. The aim of every act of the Will—or more properly speaking, of the mind,—is to secure the greatest apparent good, or what in the actual view of the mind appears to be the greatest good.

6th. The Will is not a self-determining, self-controlling power, and therefore not absolutely free ; but its freedom consists in the choice of motives.

7th. We have the power to some extent of altering the circumstances that give rise to our motives ; but in order to exercise this power effectively, and to a good end, knowledge is absolutely necessary.

I have chosen to treat this subject in a popular method, believing, as I remarked in the beginning, that it will be more useful to my class of readers than a more philosophic disquisition. Those who prefer the region of pure metaphysics are not likely to read my lucubrations, and those who prefer a more scientific treatise on the physico-metaphysical bearings of the Will have Spencer, Mill, Bain, Maudsley, and others to appeal to.

CHAPTER III.

ANATOMY OF THE SKULL.

THE skull is divisible into two parts, the cranium and the face ; the former is the bony cavity which contains and protects the brain, and is composed of eight bones :—*occipital, frontal, sphenoid,* and *ethmoid,* two *temporal,* and two *parietal.* Most of their borders are serrated, and make, when articulated, zigzag sutures.

The *temporal* and *parietal* bones (fig. 8, 4, 2). form the sides of the head ; the former being situated at the inferior portion, and the latter at the upper. The temporals contain the auditory apparatus, and are divided into three portions, —the *squamous,* the *mastoid,* and the *petrous.* The petrous portion projects inwards at the division of the middle and posterior lobes of the brain. The squamous portion is a thin plate of bone that articulates with the parietals at its upper edge, which is bevelled from the inside, and forms the squamous suture. A lengthened process proceeds from the temporal bones, projecting outward and advancing forward until it joins a similar process projecting backward from the malar bones, and the two form the zygomatic arch. The two parietal bones articulate at the median line by the sagittal suture, and are attached to the temporal bones by the squamous

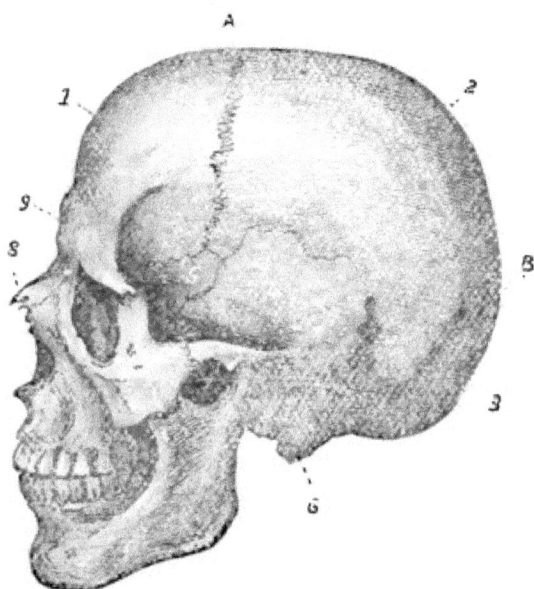

Fig. 8.—Side View of Human Skull.

1, Frontal bone ; 2, Parietal bone of the left side ; 3, Occipital bone ; 4, Temporal bone ; 5, Upper portion of the great wing of the Sphenoid bone ; 6, Mastoid process ; 7, Zygoma ; 8, Nasal bone ; A, Coronal suture ; B, Lambdoidal suture ; C, Squamous suture.

suture (fig. 8, c). They contain the crown, and descend a little backward, and advance forward over half the length of the head.

The *occipital bone* (fig. 8, 3) composes the greater portion of the base of the cranium, and unites anteriorly with the sphenoid, and laterally with the temporals.   It contains the foramen magnum, or large opening, for the passage of the spinal cord, from which part it proceeds backward to the extremity of the cerebellum, then bends itself upward, and expands laterally for a short distance.

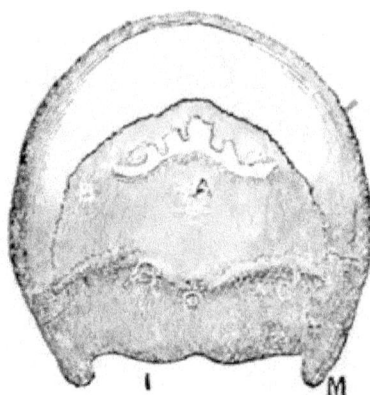

Fig. 9.—Back View of Human Skull.

A, Occipital bone ; B, Lambdoidal suture ; O, Occipital spinous process :
M, Mastoid process ; S, The transverse ridge, and the superior margin of the cerebellum ; I, Inferior margin of the cerebellum.

It afterwards greatly diminishes in size until it terminates in a point at about two-thirds of the height to the crown ; and it is attached to the posterior borders of the parietal and temporal bones by the lambdoidal suture.   The form of this bone is said by anatomists to have some resemblance to the Greek letter Λ (*lambda*, fig. 9, A), from which

circumstance its articulating suture derives its name.
It contains internally the cerebellar fossa,—the bed
of the cerebellum,—and on its external surface the
transverse ridges, occipital spine, and spinous process are situated.

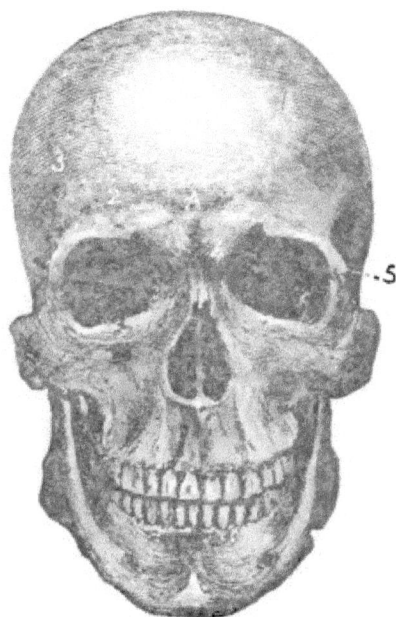

Fig. 10.—Front View of Human Skull.

1. Point of ossification of the frontal bone ; 2. Superciliary ridge : 3.
Temporal ridge ; 4, Internal angular process ; 5, External angular
process ; 6, The orbit ; A, Frontal sinus.

The *frontal bone* (fig. 8, 1) forms the forehead
and the vaults of the orbital cavities, and it is joined
to the parietals by the coronal suture.   Its upper
posterior part is not fully developed at birth, and a
space between it and the parietal bones is un-

covered with bone in consequence. This space is called the fontanelle.

There are two points of ossification in the frontal bone (fig. 10, 1) ; they are situated at the upper and outer angles of the forehead, and at the most convex parts of the bone. In addition to these points, the following should be specially attended to :—The superciliary ridge, the temporal ridge, and the internal and external angular processes (fig. 10, 2, 3, 4, 5).

The *sphenoid bone* resembles a bat with extended wings, and is situated at the anterior portion of the base of the skull. It wedges in between all the other bones, and supports them, like the key-stone of an arch. The lesser wings (fig. 8, 5) are about an inch broad, triangular in shape, and extend laterally and upwards between the frontal and temporal bones, their tips coming in contact with the parietal bones, thus making up a part of the external form of the cranium, and the anterior sides of the angle form part of the orbital cavities.

The *ethmoid* is a small sieve-like bone, which is situated at the anterior inferior part of the base of the skull, behind the nose, and is covered by the bulb of the olfactory nerve in the living head. In a phrenological point of view it is unimportant.

In the order of development the brain is formed first, and the skull is afterwards moulded to it. The skull first appears in the foetus as a gelatinous membrane, and small particles of bone are deposited on its surface at different parts, which are called the

points of ossification, and from them rays of bony particles proceed during the process of development.
The points of ossification in the frontal and parietal
bones are often referred to by phrenological writers,
and the student of phrenology should be acquainted with their situation and distinguishing features.
Each of the parietal bones is developed from one
point only, which is situated about the centre of
the bone at its most convex part, and gives it a prominent appearance. It is easily discerned in the
living head by placing the palms of the hands upon
the parietal bones from behind. If a line were
drawn from the opening of the ear up to the middle
of the crown, it would pass through the centre of
the point of ossification, or thereabout.

The skull consists of three layers; two compact
plates, also called tables; and the diploë,—a network of bony tissue between them. The two plates
are not exactly parallel to each other, but the divergence is slight—not exceeding one-sixth of an inch,
except at the frontal sinus. The parts of the skull
which are generally thickest are much the same in
all persons. The thinnest parts are at the orbital
plates and the squamous portion of the temporal
bones; and the thickest is along the median line,
especially from the top of the brow to the end of
the frontal bone, and from the crown to the foramen magnum.

The frontal sinus is a cavity caused by a separation of the plates of the skull at the root of the

nose, the inner angles of the eyes, and between the eyebrows (fig. 11, D). It is not the same size in every individual, neither is it in all cases uniform at each side of the median line in the same head, nor as to the area it covers, or the extent of divergence

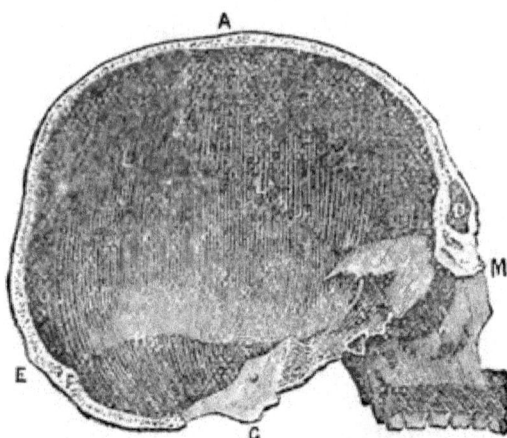

Fig. 11.—Section of a Human Skull.

Cut down through the median line, photographed, drawn, and engraved, about one-fourth the natural size, showing the exact thickness and extent of divergence from parallelism of the tables.

A, Outer table ; B, Diploë ; D, Frontal sinus ; E, External occipital spinous process ; F, Internal spinous process ; G, Mastoid process ; H, Internal transverse ridge.—the situation of tentorium ; L, Grooves for arteries ; M, Nasal bone.

of the plates. This really presents a difficulty to the phrenologist in estimating the size of the organs covered by the sinus ; and it has been repeatedly advanced as sufficient to undermine the whole science. Such an argument, if it may be so-called, only tells against the persons who advance it, as it shows them to be either deficient in knowledge of

the subject, or in intellectual capacity to grasp it;
or that they are so blinded by bias, as to unfit them
for viewing it in a proper light, and for arriving
at a legitimate conclusion.   For, supposing it were
impossible to form a correct judgment of the size
of a few organs in the neighbourhood of the sinus,
this fact would not invalidate the proofs of the re-
mainder.   It would be just as sensible to assert
that, because a blind person cannot see, he is
therefore deaf, as to say that, in consequence of
a difficulty in ascertaining the dimensions of the
organs of Form, Size, and Individuality, the size
of the organs that are situated in the moral and
social regions cannot be estimated.

The frontal sinus rarely appears before the age
of twelve.   Therefore, the best proofs of the existence and the size, and consequent power of the
above organs, are to be obtained previous to, at, or
shortly after this age.

In old age the brain is said to diminish in volume,
and in some cases of disease to a great extent, which
is not indicated by a corresponding contraction of
the head, in consequence of the inner table of the
cranium only receding with the brain by separating
from the outer table.   In such instances an accurate diagnosis of the mental powers cannot be obtained from the form of the head.

CHAPTER IV

THE mind and body are reciprocally united by the gordian-knot of life until death divides it.

All that is known, or can be known of mind, are its attributes as manifested through, or by means of, the body.

The body is the organ of the mind, but the nervous system constitutes its more immediate group of organic instruments; but phenomena which are recognized as mental, are said to be evolved in the cerebral hemispheres, hence the brain is spoken of as the organ of the mind. Some medico-psychologists, however, take exception to this designation, and contend for the inclusion of other centres.

It is not my intention to go into this question, and I shall treat of the brain as the mind's chief organ.

Where there is no brain or nerve there is no

mind, and the lowest forms of animal life possess
the most rudimentary-mind organism ; and as we
ascend the scale of mental nature, we find that parts
are added to the nervous system ; the brain is more
complex in structure, is richer in all that appertains
to the manifestation of mind, until we arrive at man,
who stands, giant-like, at the head of the animal
series in mental status, while he possesses a brain as
far above them in development and richness of
structure as he does in mind.   In order to get a
correct idea of the attributes of mind, a knowledge
of the structure and functions of the nervous system
is a useful, if not a necessary acquisition, and the
following compendium is given as an aid to the
uninitiated in this branch of study.

The nervous system consists of a grand trunk,
called the cerebro-spinal centre, which is divided into
two portions : the cerebro-centre or brain, and the
spinal-centre or spinal cord, and forty pairs of nerves
—nine pairs of these are united at their inner terminations with the brain, and are named cranial
nerves ; thirty-one with the spinal cord, named spinal
nerves.   An idea may be got of the ramifications
of their outward terminations, from the fact that no
feeling or movement can take place in the absence
of a nerve.

The brain fills the cavity of the skull, and the
spinal cord is situated in the vertical canal.

The cranial nerves are arranged in groups, and

named by Willis in accordance with their functions,
thus :—

Special Sense	1st, Olfactory. 2nd, Optic. 7th, Auditory.
Motion	3rd, Motus Oculorum. 4th, Pathetic. 6th, Abducentes. 7th, Facial. 9th, Hypoglossal.
Compound,—that is of *Motor and Sensory* ...	5th, Trifacial. 8th, Glosso-Pharyngeal. „ Pneumogastric. „ Spinal Accessory.

It is desirable to recognize the following particulars:—*First*, the elementary substance of the nerves;
second, their structure and origin ; *third*, the form
and structure of ganglions and plexus ; *fourth*, the
cerebro-spinal cord and its nerves ; and *fifth*, the
brain and the nerves which proceed from it.

The nerves are composed of a white fibrous substance (misnamed medullary), and of an ash-colored
or grey pulpy substance, which consists of nucleated
cells or corpuscles of various forms, and differing in
size from $\frac{1}{200}$ to $\frac{1}{6000}$ of an inch in diameter ; and
nuclei from $\frac{1}{1500}$ to $\frac{1}{8000}$ of an inch in diameter.
Nerves are whitish cords, made up of numerous

tubular filaments, running parallel, and varying in thickness from $\frac{1}{1200}$ to $\frac{1}{14000}$ of an inch, which are invested in a cellular membrane. If a skein of silk were covered with a layer of paper, and cut in two, it would be a very good illustration of a nerve,—the paper representing its investing membrane, and the threads of silk the fibres.

Ganglions are knots or swellings, of irregular form, and are composed of the white and grey substances, the latter being internally disposed.

A plexus is a nervous network, which is formed by the nerve filaments diverging and interlacing, or weaving one with another, then converging and re-uniting again to form a cord as before. In all these interlacements, or plexuses, there is a quantity of grey matter, as in the ganglions.

The spinal cord (fig. 12) is divided into two similar lateral halves by an anterior and posterior fissure, each half having three columns. Thirty-one nerves proceed from each side by two roots, one root proceeding from the anterior column, and the other from the posterior (C, E, fig. 12). The anterior roots give rise to the motor nerves, and the posterior roots to the sensory nerves. On each posterior root a ganglion is formed, from which proceed several fibres (D, fig. 12), then the motor and sensory roots unite and become indistinguishably blended in one common sheath, and are thus distributed to the parts of the body which they are intended to serve, sending off in their course numerous branches, and gradually lessening in size, until their terminations cannot be detected by the unassisted eye.

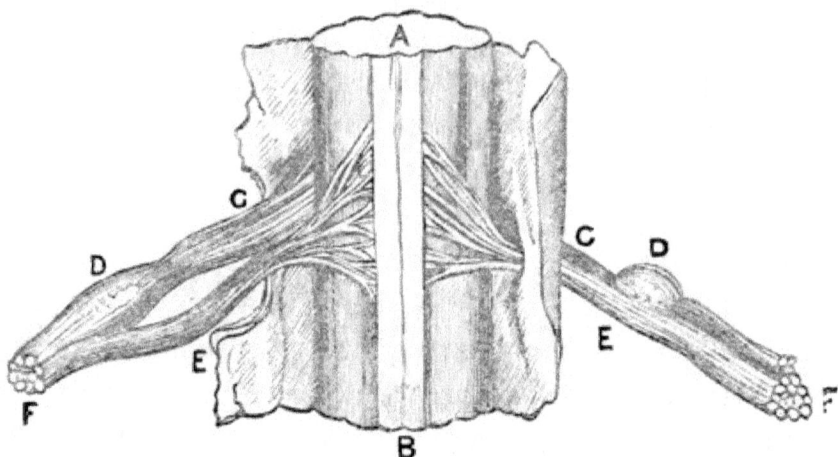

Fig. 12.—Anterior View of the Section of the Spinal Cord.

A, B, The spinal marrow divided into lateral portions ; C, The spinal sensory nerve rising from the posterior lateral division ; D, Ganglion of the sensory nerve ; E, The motor nerve which takes its rise from the anterior lateral division ; F, Union of the sensory and motor nerves. These nerves proceed to their destination blended together in one common sheath.

The brain and spinal cord are invested with three membranes. The outer one is firm and strong, and is called the *dura mater* (hard mother). It sends forth a process which dips down between the hemispheres of the brain, named the *falx cerebri*, or falciform process of the dura mater, in consequence of its great resemblance to a sickle ; it is narrow in front, broad behind, and has a sharp curved edge below, and attaches in front to a process of the

ethmoid bone, and behind to the tentorium cerebelli.
The dura mater serves as a lining to the skull, and
firmly adheres to it.   Immediately beneath the dura
mater is the *arachnoid*, a serous membrane, so named
from its resemblance to a spider's web.   The next
is the *pia mater* (soft mother); it is an areolar,
vascular membrane, composed of innumerable blood
vessels, and it dips into the sulci or furrows, and
covers the entire brain.

The encephalon or brain consists of many parts,
which are composed of white and grey nervous substances.   Each part either forms a distinct centre,
or a medium of communication between the centres.
The principal portions are the MEDULLA OBLONGATA
(long marrow), the CEREBELLUM (little or after brain),
and the CEREBRUM, or brain proper.

The *medulla oblongata* is situated at the top of the
spinal cord, and is a continuation of it.   It is rather
conical in shape, with its broad end upwards, and is
about 1·3 inch in length.   It is divided into two
symmetrical lateral columns by two vertical fissures,
the anterior and posterior median fissures, and each
column is subdivided by shallow grooves into three
smaller cords, consisting of what are called the *corpora pyramidalia*, the *corpora olivaria*, and the *corpora restiformia*.   Quain says—

"These cords, according to Gall, are made up of the primitive and formative fibres of the cerebrum and cerebellum ;
for, if they be traced upwards, the anterior pyramids and the
corpora olivaria will be found continuous with the fibres

which are expandeded into the cerebral hemispheres, whilst
the posterior pyramids (usually called *corpora restiformia*) are
evolved into the lobes of the cerebellum."—*Elements of
Anatomy*, 3rd ed., p. 684.

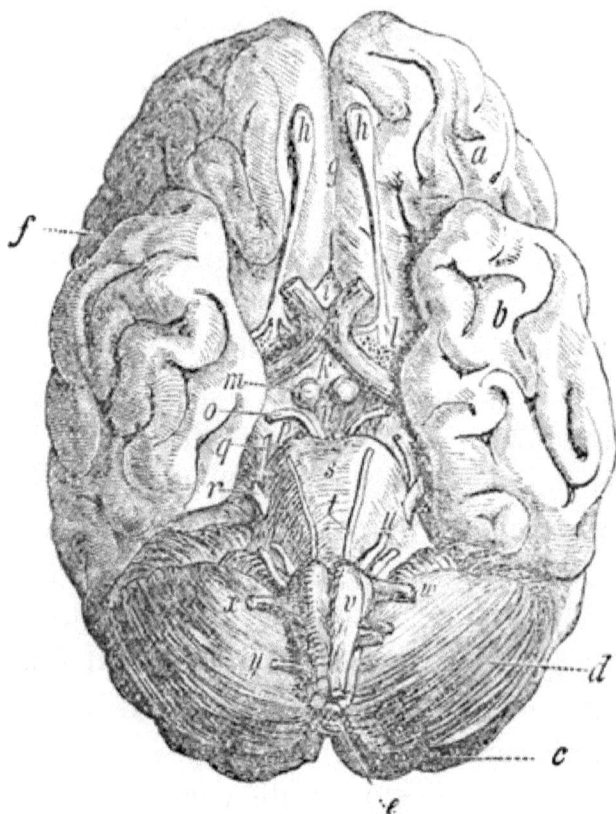

Fig. 13.—Base of the Human Adult Brain.

a. Anterior lobe of cerebrum. *b*. Middle lobe, *c*. Posterior
lobe, appearing behind the cerebellum. *d*. Hemisphere
of the cerebellum. *e*. Lower extremity of the medulla
oblongata. *f*. Fissure of Sylvius. *g*. Longitudinal
fissure. *h. h.* Olfactory bulbs. *i*. Optic commissure. *l*.
Three roots of olfactory nerve. *m*. Corpora albicantia.

v. Third pair of nerves. *p*. The crura cerebri. *q*. Fourth nerve. *r*. Fifth pair. *s*. Pons Varolii. *t*. Sixth pair of nerves. *u*. Seventh pair. *v*. Anterior view of the pyramidal bodies ; *w*. and the two below are the eighth pair ; between *w* and *v* is the olivary body. *x. y.* Two roots of ninth pair of nerves.

Fig. 13 is a representation of the base of the encephalon divested of its membranes, showing the anterior portion of the medulla oblongata, the *pyramidal bodies*, the *oliviary bodies*, and a small portion of the *restiform bodies*, the inferior surface of the cerebellum and cerebrum, the *pons Varolii*, and the *crura cerebri*. Crura (legs) are here used to denote bundles of fibres which proceed from the medulla to the grey substance of the cerebellum and cerebrum, as seen issuing from the *pons Varolii* in their passage upward (fig. 14, G, 34–38).

The *cerebellum* occupies the base of the posterior part of the cranium behind the ears. Its under surface is convex, and rests in the occipital fossa ; and the sides of its superior part ascend to a level with the transverse ridge of the occipital bone, but the centre rises a little higher towards the median line. It is divided into equal lateral hemispheres, the division being formed behind by a fissure which receives the *falx cerebri*, and in front by a deep cleft that lodges the medulla oblongata. It is composed of white and grey matter, the latter being spread over the entire surface in considerable thickness, which is folded in numerous parallel laminæ placed vertically with their edges uppermost, imparting a streaked appearance externally, and, on a section, a tree-like aspect, well branched and covered with foliage ; hence it is named *arbor vitæ* (tree of life). In each hemisphere there is a ganglion (an oval mass of grey matter) called the *corpus dentatum* (*s*, fig. 14), which, according to Gall, is a ganglion of increase to the formative fibres of the cerebellum. The two hemispheres are united by the *pons Varolii*.

"The *pons Varolii* is a comparatively small portion of the encephalon, which occupies a central position on its under surface, above and in front of the medulla oblongata, below and behind the crura cerebri, and between the middle crura of the cerebellum, with all which parts it is connected. The substance of the pons Varolii consists of transverse and longitudinal white fibres, interspersed with a quantity of diffused grey matter. The transverse fibres, with a few exceptions, enter the cerebellum, under the name of the middle peduncles, and form a commissural (or connecting) system for its two hemispheres. The longitudinal fibres are those which ascend from the medulla oblongata into the crura cerebri, augmented, it would seem, by others which rise within the pons from the grey matter scattered through it."—*Quain.*

The cerebellum is separated from the cerebrum by a strong membrane (the *tentorium*, a tent), which is attached at each side to the margin of the petrous portion of the temporal bone, and behind to the transverse ridge of the occipital bone. It supports the posterior lobes of the cerebrum, and prevents their pressure on the cerebellum. In leaping animals, the tentorium forms a bony tent. The cerebellum bears no fixed relative proportion to the cere

H

brum, but differs greatly in individuals.   In the adult, the former varies in size in the ratio of from one to six or eight of the latter ; but in infants the disproportion is greater, being, according to some. anatomists, one to thirteen or fifteen.   But, if we accept this estimate as a general rule, too much reliance should not be placed on it, for the exceptions to it are numerous.   In the autumn of 1873 and spring of 1874, I examined the heads of many children whose cerebellums were proportionately larger to their cerebrums, than I have generally found in the adult head : and several of their parents said, in answer to questions I put to them. that they had observed unusual amative indications in their demeanour.

The external surface of the cerebrum is entirely covered with the grey substance, which has two names. one being derived from its position,—cortical substance,—and the other from its colour,—cineritious substance.   This substance is disposed in the cerebrum in a series of convolutions something like a lady's cap-border, the outer parts being of a rounded or convoluted form, while the inner ones, or those which project internally, are sharp and narrow. similar to the underside of a frill.   In consequence of the superior surface being convoluted, and the convolutions lying in juxta-position, there are furrows between them, which are called by anatomists fissures and sulci.   If the student press his

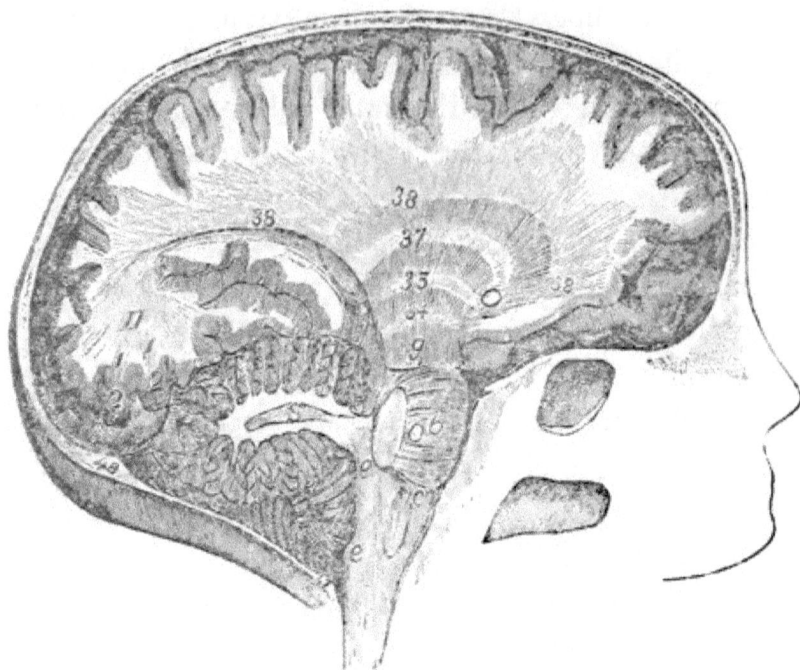

Fig. 14.—Section of Skull and Brain.

The incision is made in the right side on a line with the orbit, and passes vertically through the cerebral and cerebellar hemispheres. The two lines bounding the circumference represent two tables of the cranium, drawn proportionately wider apart than nature to show the diploë. *e, e.* Section of the corpora restiformia. *c.* Section of corpora pyramidalia. *b.* The pons Varolii. *g.* One of the crura of the cerebrum. 34, 35, 37, 38, 11. The white divergent fibres of the crura in their passage from the corpora pyramidalia upward through the pons Varolii, thalami optici, and corpora striata, which ultimately terminate in the cerebral convolutions. The dark masses between 34–38, represent grey matter; 47, 48, situations of the cerebellum within the cranium. *s.* Corpus dentatum.

fingers together, the backs of them will represent the convolutions, and the furrows between them the sulci.

I may here state that, as the comparative terms, superior and inferior, will be often used in describing the convolutions and sulci, they are used in anatomy in reference to position, as upper and lower, not to quality, as better and worse.

EXTERNAL FORM OF THE CEREBRUM.—The cerebrum is composed of two hemispheres, which are united together by transverse medullary fibres, which will hereafter be described. Its external surface presents to the eye numerous irregular convolutions, separated by fissures or sulci, which were considered by the ancients to be accidentally disposed, and little attempt, if any, was made to discover method in their arrangement, until comparatively recent times, when the attention of De Vicq-d'Azyr, Sœmmering, Gall, Spurzheim, Vimont, Rolando, Foville, and others, was directed to the unravelment of the apparently inextricable labyrinth of the convolutions, which seemed to have neither beginning nor ending, nor design. These close observers did much to bring order out of chaos. And at a subsequent stage, Leuret threw his energies into this field of inquiry, and well-nigh succeeded in the discovery of methodical arrangement in the disposition of the convolutions, when early death called him from his labours, and it fell to Gratiolet to complete the cerebral topography, which work he appears to have finished successfully.

We have not been without earnest and painstaking inquirers at this side of the Channel, among whom Professors Huxley and Turner take the foremost rank.

Dr. Ecker, Professor of Anatomy in the University of Freiburg, Baden, has likewise contributed largely to elucidate the subject.

The convolutions of the human brain are found to be arranged on a general plan, although there are considerable minor diversities in the brains of different persons, and even the two hemispheres in the same individual are seldom symmetrical. The convolutions of some brains are very tortuous and complicated, while in others they are simple, and their course is much more easily traced. The former state exists in persons of a higher intellectual order, than individuals who have brains of the latter kind.

In the early stages of fœtal development, the external surface of the cerebrum is smooth, and, excepting the Sylvian fissure, no sulci are observed until about the fifth month, when the fissure of Rolando makes its appearance, and others follow in succession as the fœtus develops.

The convolutions and sulci are best distinguished in the more simple structural forms; and the order of their development should be studied in the fœtus at the various stages of its age and growth. Ecker

considers it above all things necessary to examine
numerous brains of embryos, in order to get a clear
view of the cerebral architecture, and to distinguish
what is essential from what is accessory. Gratiolet
began his researches by studying the cerebral forms
of the simiæ, and he commends this method.

Plate 1 is an engraving of the left hemisphere
of an adult human brain. It was selected out of a
large number, as being moderately rich in convolutions, and an excellent representative specimen.

The superficial surface of the cerebral hemispheres is convex, but the median surface is flat.
Each hemisphere is divided into five lobes, four of
which come in contact with the skull, and are
marked off by fissures, more or less defined. They
are named the *frontal, parietal, tempero-sphenoidal,
and occipital lobes.* The other one does not appear
on the external surface, and is hid from view. It
is called the Island of Reil or central lobe.

Let us, now, by the aid of the engraving (plate 1)
trace out these lobes and the principal convolutions of each. The best mode of procedure which
suggests itself is that which we would be likely to
adopt in studying the plan of a town,—that is, to
observe the chief thoroughfares first, and take the
principal buildings as guide-posts, for the better
comprehending and remembering of the secondary
streets and minor objects of interest.

The sulci are divisible into three orders :—*Pri-*

mary,—the leading thoroughfares; *secondary* and *tertiary;* and the lobes into primary and secondary or lobules, and gyri or convolutions.

THE FISSURE OF SYLVIUS.—The lowest portion of the cerebral hemisphere, it will be seen, is in the centre, at the anterior terminus of which, and about one-fourth way from the frontal extremity of the cerebrum, there is a deep indentation,—this is the fissure of *Sylvius* (F. S.). It has two rami or branches. A short one,—the frontal ascending ramus (A. S.),—that ascends, and is circumscribed by a fold of the infero-frontal convolution; and a longer one,—the horizontal ramus (H. S.),—that stretches backward considerably, then sends off another ascending branch into the supra-marginal gyrus. This fissure makes its appearance at the third month of fœtal life, and is different in structure from the rest; for, whereas these are formed by the folding of the cortical, it forms a distinct cleft internally. It originates behind the roots of the olfactory nerve, and, where it appears at the cortex, takes a form like the letter Y, by sending off the rami just named.

The FISSURE OF ROLANDO,—called by Huxley the *postero-parietal* sulcus, and by Ecker, in common with some of his predecessors, *sulcus centralis.* Two convolutions (F. A., P. A.) may be observed ascending obliquely backwards from the Sylvian fissure up to the median edge of the hemisphere; and that they are divided by a deep fissure (R), excepting at their superior and inferior borders. This is the fissure of Rolando. It appears in the fœtus at or about the end of the fifth month.

The Parieto-Occipital Fissure (P. O.) projects from the median margin in a transverse direction, immediately behind the superior parietal lobule, and in front of the superior occipital gyri. It is generally short superficially, and, in some cases, appears as a mere notch in the median margin; but Professor Turner has seen it more than two inches long in the right hemisphere of a female.* It is more constantly distinct in the median surface, and is said to show itself at the end of the fourth or beginning of the fifth month. Gratiolet divides it into two, and he names them respectively the internal and external perpendicular fissures.

The Intra-Parietal Fissure. — This fissure (I. P.) exists in the brain of the ape, as well as in the brain of man; yet, according to Ecker, it has only been described as typical by Pansch and Turner. It generally arises above the horizontal ramus of the Sylvian fissure, and ascends between the parieto-ascending and the supra-marginal convolutions to the top of the latter, then turns backward, and after

* Edinburgh Medical Journal, No. 132, p. 1111

passing through the parietal lobe, dividing it into superior and inferior lobules, bends in a lateral and downward direction, and runs for a variable distance nearly parallel to the median fissure.

The fissures of Sylvius and Rolando mark off the anterior lobe, and the anterior-inferior boundary of the parietal lobe. The tempero-sphenoidal lobe is bounded at its superior and anterior margins by the horizontal ramus of the Sylvian fissure; and the parieto-occipital fissure distinctly divides the occipital lobe from the parietal lobe at the median surface and partially at the vertex.

We have now traversed the boundaries of the lobes, and are in a position to comprehend the situations and forms of four of them, but the central lobe being hid from view, its locality can only be pointed out. This lobe is roofed in by those convolutions which cluster round the Sylvian fissure, where it forms something like the letter Y.

According to Gratiolet, the Island of Reil in the human brain has five or six radiating gyri, but it is smooth in the lower mammals. Some anatomists say that, in the orang-outang and chimpanzee, one or two sulci exist, indicating a convolved surface; but Gratiolet thinks this requires confirmation.

The FRONTAL LOBE (FR.) is that portion of the cerebrum which lies in the orbital cavity, and consists of the frontal ascending convolution (F. A.), directly in front of the fissure of Rolando, and three horizontal convolutions that lie parallel to each other,—the supero-frontal (s. F.), the medio-frontal (M. F.), and the infero-frontal convolutions (I. F.). They jut out of the frontal ascending convolution ; in fact, this convolution, as a general rule, curves round at each end, and runs forward continuously with the supero-frontal and infero-frontal convolutions. The middle one, in many cases, is cut off superficially from the ascending gyri by the anterior ascending sulcus (d). These convolutions advance forward to the extremity of the hemisphere, then bend round, downwards, and return backward against themselves to the anterior ascending convolution.* They are divided on the superior surface by two sulci,—the supero-frontal sulcus (f), and the infero-frontal sulcus (g), which are difficult to trace in some brains, in consequence of the complicated windings of the convolutions. Besides, processes frequently project from the convolutions across the sulci, and so displace them as to bewilder the eye in tracing their course. I have a cast of Dr. Andrew Combe's brain, and such is the tortuous complication of the convolutions, that even the primary sulci are difficult to trace.

Gratiolet and other French anatomists, with whom the English anatomists agree, divide the frontal lobe

---

* Observe, the convolutions as they appear are only described here, not the order of development.

into two lobules, the one just described, and a smaller one which lies in the cavity of the frontal bone,—the orbital lobule; but Ecker disputes this, and considers the frontal lobe as being indivisible into lobules, and treats of it as such.

The TEMPERO-SPHENOIDAL LOBE (T. S.) lies at the base of the middle portion of the skull in the temporal fosse.   Its anterior and superior borders are distinctly marked by the Sylvian fissure, but the posterior boundary is ill-defined; and, it not uncommonly merges into the parietal and occipital lobes without any line of demarcation.   It has three convolutions that lie parallel to each other (S.T.S., M.T.S., I.T.S).

The PARIETAL LOBE (P.A.) is situated between the frontal and occipital lobes, and above the tempero-sphenoidal lobe.   It is divided into a superior and inferior lobule by the intra-parietal fissure (I. P.) and consists of four convolutions,— the parieto-ascending (P. A.), superior parietal lobule (S. P. L.), the supra-marginal (S. M.), and angular convolutions (A. G.).   The superior lobule issues from the upper hind part of the ascending convolution, and extends back to the parieto-occipital fissure.   The supra-marginal gyrus comes out of the lower end of the ascending convolution posteriorly, and ascends to the horizontal branch of the intra-parietal fissure, then bends downward and

passes into the upper tempero-sphenoidal convolution.   The angular gyrus arises out of, and
lies behind the supra-marginal gyrus, and unites
inferiorly with the middle tempero-sphenoidal convolution.

The OCCIPITAL LOBE (O. C.) is the smallest in the
human brain.   Its superior anterior border is divided
from the parietal lobe by the parieto-occipital fissure ;
but excepting this, it has no definite anterior border
line.   It has three tiers of gyri,—superior (1), middle
(2), and inferior (3).   The inferior gyrus runs into
the inferior gyrus of the tempero-sphenoidal lobe.

With regard to the convolutions of the median
surface of the hemispheres, little need be said.

The *gyrus fornicatus* (G. F. plate 2) arises under
the anterior fold (genu, c) of the *corpus callosum* by
a narrow strip, and gradually increases in size as it
circumscribes this commissure, and after bending
round its posterior fold (the *splenium*, d) it fuses
into another convolution called the *hippocampi*.   The
superior frontal convolution (S. F.) encircles the anterior and superior portions of the *gyrus fornicatus.*
These convolutions are divided by the *calloso-mar-
ginal fissure*, which also forms the anterior boundary
of the *quadrate lobule* (Q. L.), while the inferior parieto-occipital fissure forms the posterior boundary
of this lobule, and the anterior border of the *cuneus*
(Q.),—a wedge-shaped convolution,—its inferior border being marked off by the *calcarine fissure.*

Anatomists describe the outer surface of each cerebral hemisphere as being divided by fissures into four lobes, notwithstanding all the lobes are more or less superficially united ; and it is worthy of note, that the proportion each lobe bears to the others is not invariable, but greatly differs in different persons : a fact that has an important bearing in practical phrenology. Two more facts are equally noteworthy,—namely, that the courses of the fissures of Rolando and Sylvius likewise vary in various brains.

Professor Turner says, " The more highly developed the frontal lobe of the brain is, the more oblique is the fissure of Rolando."—*Edinburgh Medical Journal*, No. 132, p. 1110. It will be observed that Professor Turner not only recognises variableness in the obliquity of this fissure, but he accounts for it. In his opinion, increased obliquity is caused by increased development of the frontal lobe ; and Professor Ecker expresses himself in like manner.—*Convolutions of Human Brain*, pp. 11, 12. These anatomists are known as painstaking and accurate observers, and I am especially disinclined to question the accuracy of any conclusion to which they have arrived of a purely anatomical nature : notwithstanding, I think they err in describing the cause of great obliquity of the fissure of Rolando to great development of the frontal lobe. This will probably be the effect in cases where the upper-posterior portions of the superior and medio-frontal

gyri are largely developed, so as to extend considerably backward; but an opposite result would be
produced by the bulk being added to the anterior
and coronal portions of these gyri; and, moreover,
the frontal lobe might be increased laterally without
adding to the obliqueness of the fissure in question;
but, not only so, this might produce an opposite result.

Professor Turner, as I shall explain by and by,
has observed a relation between the course of the
fissure of Rolando and that of the coronal suture of
the skull: that the former is behind the latter about
1·2 or 1·3 inch at its lower end, and 1·5 inch at
its upper end. Now, accepting this as the general
rule, we have only to find the situation and course
of this suture to determine the locality and disposition of the aforesaid fissure.

Having examined some hundreds of skulls, with
the aim of testing the accuracy of the conclusion
of Professors Turner and Ecker with regard to the
cause of great obliqueness of the fissure of Rolando,
I find that the coronal suture more nearly ap-
proaches the perpendicular in skulls that are more
highly developed in the upper frontal and coronal
regions, than it is in such as have these regions
depressed; and, I have further observed, that large
lateral development of the frontal bone and a comparatively vertical coronal suture are frequent concomitants. And Dr. J. B. Davis, the distinguished
ethnologist, unhesitatingly corroborates my observations in a letter dated Jan. 17, 1875, which I received
from him in answer to my inquiries on the subject.

There is an engraving of the left hemisphere of
an adult human brain in Dr. Vimont's splendid
Atlas, consisting of 800 subjects in comparative
anatomy, and the development of the infero-frontal
gyrus is so uncommonly large posteriorly as to cause
the fissure of Rolando to incline considerably forward from below upwards. The inferior margin of
the fissure of Rolando is actually displaced backward
beyond the indentation that is caused by the petrous portion of the temporal bone, and the frontal
ascending gyrus completely overlaps the tempero-
sphenoidal lobe, and descends to the basal margin
of the hemisphere at the indentation just named.
If this be a correct drawing, it represents a most
remarkable brain in every respect ; for the frontal
lobe is larger than all the others put together.
Should there be any similar brains in the skulls of
living human beings, I should like to see them, and
to know their history ; but, I candidly confess, I
should not undertake to read their characters ; notwithstanding, I think great talent would be palpably
indicated.

A question suggests itself here. Are there any
outward indications in the head of the relative size
of the lobes of the cerebrum that come in contact
with the skull, and by which the extent of obliqueness of the fissure of Rolando and the ascending gyri
may be inferred? or whether or not they be oblique
in any particular head? I think there are such ;
and I shall endeavour to point them out when I

come to treat of the rules for interpreting the signs
of character indicated by cranial conformation.

On parting the hemispheres a little at their upper
surface with the fingers, a broad band of white substance (*corpus callosum*) is seen to connect them.
The corpus callosum is a thick layer of medullary
fibres passing transversely between the two hemispheres, and constituting their *great commissure.*
It is about four inches long, and is situated nearer to
the anterior than to the posterior margin of the brain.
There are, also, the anterior and posterior commissures which assist in uniting the hemispheres.   If
a superficial incision be made through the corpus
callosum on either side of the *raphé* (seam) at the
median line, two irregular cavities will be opened,
which extend from one extremity of the hemispheres
to the other ; these are the lateral ventricles, from
which may be seen the superior surfaces of the
corpora striata and *thalami optici* (the superior and
inferior cerebral ganglions).

The *corpora striata* are situated within the white
matter of the anterior lobe.   They are pyriform in
shape, and their external surface is composed of
grey matter, but internally the grey and white substances are intermixed, producing a striated aspect,
from which the name striated bodies is derived.
Immediately behind these are the *thalami optici,*
two ovoid masses of grey and white substance embedded in the middle lobes ; behind them, and below the posterior part of the corpus callosum and

the cerebral peduncles, four small rounded eminences project, composed of white matter without
and grey within, named *corpora quadrigemina.*

The brain is an aggregate mass of distinguishable
parts, each performing its own function, but all
being necessary for the complex manifestations of
mind, for which purpose they are all united ; and
the medulla oblongata is the grand junction between the cerebro-spinal centres.

Bundles of nerve fibres (diverging fibres) arise from
the pyramidal and olivary bodies, which ascend and
pass through the cerebral crura, the pons varolii.
the optic thalamus, and the corpora striata, diverging and increasing greatly in bulk in their passage
through each, so as ultimately to form the cerebral
hemispheres. Those arising from the corpora pyramidalia constitute the frontal lobe ; the other lobes
and cerebellum are formed of the fibres which are
sent off by the corpora olivaria and corpora restiformia. See fig. 14. Another order of fibres (converging) issue at the peripheral terminations of the
diverging ones, and proceed to the median line.
thence passing from one hemisphere to the other,
thus bringing them into relation, and forming the
commissures of the brain.

The *sympathetic* or ganglionic system is composed of a series of ganglions, united to a cord
which is situated at each side of the spinal column.
and gives off fibres that run into it, and also sends

branches to the viscera, by which these parts are brought into union with the medulla oblongata, so as to establish a sympathetic relation between the different visceral organs.

We have in the anatomy and functions of the brain and nerves a complete system of communication between the mind and every part of the body, and with the external world through the medium of the senses.

The sensory nerves, like sentinels, are posted all over to give warning of the approach of danger. The nerves of special sense receive impressions from external objects, adapted to their constitution, and transmit them inwards to certain parts of the brain fitted to receive them, where they become sensations, and from whence they are transmitted to the perceptive and reflective faculties, and there give rise to perceptions. Hence objects are seen, sounds heard, odours smelt, and flavours tasted. The other nerves of this class are designated nerves of common sensation; they are ramified over the entire body, to give the mind due notice of the bodily requirements, impending dangers, inconveniences felt, and injuries sustained, and they are the media of numerous pleasurable and painful feelings.

The nerves of motion are instruments which convey from the brain to the muscles the stimulus or nervo-vital force which incites them to action. They are divided into voluntary and involuntary. The voluntary are under the control of the will, and

obey the mandates of our volition,—as in walk-
ing, talking, eating, drinking, and writing. The
other set of organs perform their functions involun-
tarily ; thus respiration goes on, the heart pulsates,
the blood circulates, and the liver secretes bile,
whether we will or not ; hence these organs are
regularly supplied with their appropriate stimulus
through the involuntary nerves at all times, whether
we sleep or wake.

CHAPTER V.

THE RELATIONS OF THE OUTER CEREBRAL CONVOLUTIONS TO THE SKULL.

A TOLERABLE knowledge of the outer surface of the brain may be acquired by studying accurate casts and drawings of it ; but it should be observed in the skull itself to clearly understand the relations of the one to the other. This, unfortunately, is attended with great difficulty to the non-medical student ; and those who wish to acquaint themselves with the relations of the brain to the skull and head have to fall back on drawings. But although there are in anatomical works illustrations of the brain in its position in the skull, nature is seldom accurately represented. For, as Professor Ecker says :—" Men were wont to regard the convolutions as a series of folds without order or arrangement, and draughtsmen represented them much as they would a dishful of maccaroni."—*The Convolutions of the Human Brain*, p. 3.

Happily this state of things is passing away.   The topography of the convolutions and their relations to the skull, as well as the several parts they play in the mental economy, are now being taught by anatomists and physiologists of the foremost rank.

Professor Turner, of the University of Edinburgh, has signalised himself as one of the vanguard in this department.   He divides each lateral half of the cranium into ten regions, and describes the situation of the principal sulci and the convolutions of the male human brain in relation to the outer surface of these regions in a brief treatise contributed to the *Journal of Anatomy*, vol. viii. pp. 142–8 ; and in a subsequent contribution he gives an engraving on wood in illustration of them, which he has kindly permitted me to reproduce.   Plate 3 is a copy of it.

In dividing the cranium the Professor utilizes the following prominent features as guide-posts and boundaries :—The frontal and parietal eminences, or points of ossification of these bones, plate No. 4, *a, b,* the external angular process ; S. temporal ridge ; D. occipital protuberance, or spinous process, F. ; and the superior curved line, E. ; also the sutures, *g, h, i, k,* N, M ; and the squamous portion of the occipital bone.   The student should fix these structures in his memory, so as to be able to recall them before his mental vision at will before proceeding further.   Then trace the following principal regions, and afterwards the subdivisions :—The

frontal, or præ-coronal ; parietal ; tempero-sphenoid :
and occipital regions.   The coronal suture divides
the frontal from the parietal; and the tempero-
sphenoid and parietal regions are divided by the
parieto-sphenoid, the squamous, and the lambdoidal
sutures.   The squamous portion of the occipital
bone, superior curved line and occipital protuberance bound the occipital region.   This region is
not subdivided, but the others are.   The tempero-
sphenoid region is divided into a squamoso-sphenoid
and an ali-sphenoid area. See engraving. The parietal region is divided into four areas, thus :—First,
by a vertical line drawn from the posterior superior
portion of the squamous suture through the parietal
eminence up to the sagittal suture.   This makes an
anterior and posterior area, which are divided into
upper and lower areas by the temporal ridge.   The
four are named supero-anterior, infero-anterior parietal, or post-coronal areas ; supero-posterior, and infero-posterior, or præ-lambdoidal areas. The temporal ridge divides the frontal region also, into superior
and inferior areas ; and the superior one is subdivided by a line drawn parallel to the temporal
ridge, beginning at the upper portion of the orbit
and running through the frontal eminence backward
, to the coronal suture ; thus making upper, middle,
and lower frontal areas.

How to determine the Cranial Regions in
the Living Head.—An anatomist would easily de-

termine the boundaries of these regions; but persons only moderately acquainted with the human skull would not be able to do so. For the guidance of such investigators, I humbly suggest the following method, while recommending frequent manipulation of heads to find out the natural boundaries.

The coronal suture arises about an inch behind the external angular process, or the outer termination of the eyebrows, and runs upwards and obliquely backwards to the sagittal suture. This termination is easily distinguished. It is called the fontanelle—the space that is uncovered by bone in new born infants, and is vulgarly called the opening of the head.

A vertical line drawn from the centre of the hole of the ear externally over the most prominent part of the parietal eminence to the sagittal suture intersects the point in the squamous suture where the posterior boundary of the antero-parietal area begins at about one-third its length from the ear, or, roughly speaking, $2\frac{1}{2}$ to $2\frac{3}{4}$ inches upward.

A horizontal plane drawn from the external angular process, backward, so as to intersect the vertical line about $2\frac{1}{2}$ inches above the centre of the orifice of the ear, will be found to run nearly parallel to the boundary of the parietal and sphenoidal regions.

The temporal ridge arises from the external angular process, and describes an upward and backward course in an arch form, until it reaches a point immediately above the opening of the ear, and a little over half the distance from the ear to the sagittal suture, then it begins to descend, passing through the parietal eminence, and continuing backward for about seven-tenths of an inch, then acutely curves downward and forward, and terminates behind the ear in a line almost with the zygoma. In fact, the external angular process, temporal ridge and zygoma, form a pretty well-defined oval figure.

The temporal ridge is easily determined from its origin along half its course; and, in some heads, along its entire length; but it differs very much in prominence in individuals. It is sharp and well-defined in some, much broader and less prominent in others, and in particular cases its posterior half is undistinguishable in the head.

The coronal squamous and lambdoidal sutures may be distinguished in most heads after a little practice; but the attachments of the great wing of the sphenoid—the squamoso-sphenoid, parieto-sphenoid, and fronto-sphenoid—are not so easily recognised. A portion of this wing forms a part of the anterior border of the orbit. It is about an inch broad at its lower part, and from one-tenth to two-tenths broader at the upper part. The upper boundary of it is partly attached to the frontal and parietal bones, and the situation of these attachments is fairly marked off by the horizontal line that I have previously suggested as a boundary line between the tempero-sphenoid and parietal regions.

In determining the situations of the convolutions
or the parts of them within the areas, Professor
Turner saws the bone out of them, or a portion
of it, and notes the cerebral mass underlying each
in succession.   Plate 3 represents the brain in its
situation in the skull of an adult male subject,
which was completely hardened before it was uncovered.   The process of hardening causes a discharge of the fluids, and the brain to shrink.   This
accounts for the space between the skull and the
brain as shown in the plate.

It will be observed that the sutures, the temporal
ridge, and other boundary lines of the cranial areas,
are engraved on the cerebral hemisphere, in the relative positions they bear to each other.   The Professor does not state as much, but, I presume, due
allowance is made for the shrinking of the brain, in
laying down the situations of the cranial sutures
on it.

The frontal, parietal, tempero-sphenoid and occipital lobes, take their names after the cranial
bones, but these lobes are not wholly situated
within the boundaries of the respective bones from
which they derive their names.   The frontal lobe
extends behind the coronal suture under the parietal
bones considerably,—in some cases, according to
Professor Turner's observations, as far as two inches
at its upper part, and 1½ inch at the lower part;
but he puts the mean backward distance of the
fissure of Rolando (which is the posterior boundary

of this lobe) from the coronal suture to be 1·5 inch
at its upper, and 1·2 or 1·3 at its lower end.

The parietal bones likewise cover about one-
third of the posterior portion of the tempero-
sphenoid lobe, and a great part of the anterior
superior portion of the occipital lobe.

Let us now survey the boundaries of the cerebral
sulci and gyri, and distinguish the parts that lie
within each area.

The fissure of Sylvius, as explained in the last
section, divides the posterior of the frontal lobe and
the anterior of the parietal from the tempero-
sphenoid. The gyri of the last two lobes commonly
blend with those of the occipital at their posterior
terminations without any outward definite lines of
demarcation.

The fissure of Rolando, as aforesaid, forms the
posterior boundary of the frontal lobe and the anterior boundary of the parietal lobe.

Beginning at the anterior border of the tempero-
sphenoid lobe, we find that about half-an-inch of
the upper and middle gyri, with a trace of the lower
one, appears in the ali-sphenoid area (a. s.), and they
wholly occupy the squamoso-sphenoid area (s. q.),
excepting a narrow space under the squamous plate.
The middle and lower gyri, with a small portion of
the upper one, occupy the greater part of the lower
postero-parietal area (I.P.P.), besides under-lapping
its posterior boundary, and projecting a little into

the upper postero-parietal area (S.P.P.).   A triangular shaped bit of the superior gyrus, about an inch long at its upper border, appears in the lower antero-parietal area (I. A. P.) at its posterior angle, and a lesser bit of this gyrus runs into the inferior postero-parietal area.

The anterior lobe fully occupies the frontal or præ-coronal region, and nearly two-thirds of the superior and infero-antero-parietal areas, and a narrow fragment of the inferior gyrus lies under the squamous plate.   This lobe is divided into three areas, corresponding with the upper, middle, and lower gyri (S.F. M.F. I.F.).

The parietal lobe consists of two lobules,—an upper posterior, and a lower anterior, which are divided by the intra-parietal fissure, and four convolutions,—the postero-ascending gyrus, the upper parietal lobule, the supra-marginal, and the angular gyri.   The upper-parietal lobule arises from the upper posterior portion of the ascending gyrus, and extends backward by the longitudinal fissure or median line to the parieto-occipital fissure.   According to Professor Turner, the supra-marginal convolution and the parietal eminence of the skull hold a pretty constant relation, so as to warrant him in naming this gyrus, the parietal-eminence convolution.   It arises out of the lower portion of the parieto-ascending convolution posteriorly; and, behind it, the angular gyrus is situated.   About one-third of the parieto-ascending convolution, and a bit

of the parietal-eminence convolution, are located in
the hinder part of the infero-antero-parietal area.
The supero-antero-parietal area is occupied by the
remaining two-thirds of the ascending convolution,
excepting a little piece of the upper-posterior parietal lobule lying next the sagittal suture or median
line.   The parietal-eminence convolution partly occupies three areas : the one just described ; also the
anterior part of the lower postero-parietal area ; and
the upper postero-parietal area.   The lower part
of the angular gyrus stretches downward and forward into the upper posterior angle of the lower
postero-parietal area, but the greater part of it is
situated in the upper postero-parietal area at the
point of junction with the occipital lobe.

The upper postero-parietal lobule, excepting a bit
of its anterior border, is wholly situated in the upper
postero-parietal area, at the sagittal suture, extend-
ing laterally to the intra-parietal fissure (I. P.) and
backward to the parieto-occipital fissure (P. O.).
This fissure, Professor Turner says, is situated, in
general, about seven or eight-tenths of an inch
above the apex of the occipital bone, although it
varies in individuals, partly owing to a difference in
the form of the brain, and partly to the structure of
the occiput.

In examining the engraving, Plate 3, it may seem
to those unaccustomed to such representations, that
it is an error to describe the ascending convolutions

as having two-thirds of their height above the temporal ridge, and one-third below it ; seeing that the greater masses of these convolutions appear in the engraving as being below this ridge. This apparent discrepancy arises from the perspective of the drawing. The actual measurement is upward from where the temporal ridge is situated and over the top of the brain to the median line.

The vaulted part of the temporal ridge, immediately above the external meatus, is about midway between the centre of this orifice, and the sagittal suture in the skull (Plate 4), but it varies in individual cases. According to numerous measurements, I find that the average length of a line drawn from the orifice of the ear, to the sagittal suture, is about 6¾ inches, and that the temporal ridge lies a little higher than half-way ; but half-way up will express its position near enough for general practice.

Professor Turner promises a more extended memoir by-and-by, in which he will note individual peculiarities, and the difference in the relations of the skull to the brain in the sexes. I shall look forward to its publication with great interest. His cranio-cerebral map cannot fail to be of service in medical practice ; and I think it capable of being turned to useful account in practical phrenology.

The Professor has observed that the frontal eminence overlies the supero-frontal sulcus, or a tertiary convolution that crosses it in some cases. I should like to know whether or not this relation is general, or merely casual; for there is hardly a part of the human cranium in which greater individual differences exist than in the situation of the frontal eminence.

I sent home on two seperate occasions for a skull while preparing this section, but without any reference to the point in question, yet I find that the frontal eminence of the one is o·7 of an inch farther from the median line laterally than it is in the other; although the latter is by far the wider skull generally, and particularly it is o.4 of an inch wider at the parietal eminence.

Now, having marked off Professor Turner's divisions in accordance with his instructions on these skulls, I observe a great diversity in the relative size of the upper and middle frontal areas. In the narrower skull, the upper areas are nearly double the width of the middle areas, whereas the upper ones are the narrowest in the wider skull, in which the frontal eminence is closer to the median line.

Professor Ecker says that the middle frontal convolution is generally the largest, but if the relation pointed out between the supero-frontal sulcus and the frontal eminence by Professor Turner is constant, I have seen hundreds of heads in which the superior convolution would be the largest as indicated by the space between the frontal eminence and the median line.

INCREASE OF THE CRANIO-CEREBRAL DIVISIONS.
--It occurred to me while studying Professor Turner's divisions of the cranium, that they might be
advantageously increased to seventeen ; and having
had the privilege of examining a number of adult
human brains and crania with this view, in the Mus-
eums of the Universities of Edinburgh and Glasgow,
as well as of manipulating several heads during my
travels and carefully measuring them, I humbly suggest the following supplementary method.

It is impossible to lay down straight lines to
more than approximately determine the tortuous
windings of the sulci ; nevertheless, the situations of
the primary fissures may be fairly indicated.

Presuming that Professor Turner's illustration of
the brain in its situation in the skull is a fair representative specimen, a line drawn, or a piece of thin
cord stretched, across a skull or head, from the anterior border of the fosse (plate 4, x) in the zygoma
in which the head of the ramus of the lower jaw
moves, obliquely upward to the sagittal suture so as
to pass a little in front of the most prominent part
of the parietal eminence will be found to prescribe
the posterior boundary of the parieto-ascending convolution, and sever it from the upper parietal lobule
and the parietal-eminence gyrus (supra-marginal
gyrus. I shall call it the inter-parietal boundary line
ib. 4, z. B.). In cases where the squamous suture cannot be felt, about 2¾ inches from the starting point
of this line will express its position sufficiently near.

Second, a line drawn from the transverse suture (*ibid* s.), which unites the external angular process with the malar or cheek bone, backward over the squamous plate to a little behind the inter-parietal boundary line, will nearly overlie the Sylvian fissure.

Third, stretch the cord again half an inch or a little over, in front of, and parallel to, the inter-parietal boundary, from the sagittal suture nearly to the squamous plate, in order to approximately indicate the position of the fissure of Rolando, which is about 1·2 inches at its lower end, and 1·5 inches at its upper end, posterior to the coronal suture, as a general rule. I name this (*ib*. R.) the fronto-parietal boundary.

Fourth, mark off 0·8 of an inch above the apex of the occipital bone (x), at the sagittal suture to find the situation of the parieto-occipital fissure (*ib*. P.O.), and draw a line from it downwards parallel to the lambdoidal suture (E.), but diverge it forward a little the last third of the distance in approaching the squamous suture. This line will be found to nearly prescribe the anterior boundary of the occipital lobe, and may be named the parieto-occipital boundary.

Fifth, extend the line that divides the superior and middle frontal areas of the skull backward to the parieto-occipital boundary, keeping it parallel to the temporal ridge : this will divide the parietal lobules and the ascending convolutions. The anterior inferior portion of the upper parietal lobule may be found to extend a little below this line, and the upper posterior angle of the angular gyrus may pass above it a little.

We have now got seventeen areas. Three sphenoid, namely—ali-squamoso and post-sphenoid : six frontal—three præ-coronal and three post-coronal : six parietal, namely—superior, middle, and inferior parieto-ascending areas ; a supero-lobule area, upper and lower infero-parietal lobule areas : two occipital —anterior or præ-lambdoidal, and posterior or post-lambdoidal areas. The inferior posterior portion of the curve of the temporal ridge, the squamous suture and the parieto-occipital line bound the post-sphenoid area.

The situations of the convolutions are approximated by the boundary lines of the cranial areas : and the only additional descriptive remarks that it appears necessary to make, are that a bit of the posterior part of the upper *temporo-sphenoid* convolution will be found in the antero-inferior angle of the infero-parietal lobule area ; and a little of the middle and inferior tempero-sphenoid convolutions may run into the anterior occipital area.

CHAPTER VI.

SIZE OF BRAIN A MEASURE OF POWER.

IT is now generally admitted in scientific circles that there is a connection between largeness of brain and mental power. Proof of this, if any were needed, is everywhere present. We need not trouble ourselves with studying racial types and their history in search of evidence in support of it. We have merely to open our eyes to see the fact and to convince us of its truth, for it stands out in unmistakeable prominence.

Individuals of all races that have distinguished themselves by stepping out of the common rut, and have left their impress in history, have had large brains.

The average weight of the European male brain is about 48 oz., and that of the female 44 oz. The brain of Cuvier weighed 64 oz., Dr. Abercrombie's 63 oz. Bonaparte, Franklin, Daniel Webster, and Daniel O'Connell had also large brains, and were marked for corresponding intellectual power.

The brains of idiots, on the contrary, are generally small. Several have been weighed, and found to vary from 27 to 8½ oz. However, there are other causes of idiocy; and a small brain cannot be legitimately predicated of every idiot. The form of an average sized brain may indicate unmistakeable signs of idiocy; and these marks may be absent in an idiot whose head is moderately developed. And, moreover, a large head is not an infallible and constant sign of a vigorous intellect. The position of the cerebral mass, or the relative size of the lobes, indicates the measure of intellect more than absolute volume. A massive intellect and a head of moderate dimensions are not rare concomitants; Mr. Grote is said to have been a notable example.

QUALITY OF BRAIN is likewise a measure of power. This fact has forced itself on the attention of medico-psychologists; and few, if any, would attempt to gainsay it.

Persons having heads of like size and form, do not possess equal mental power if the quality of their brains be dissimilar. In fact, small headed individuals, in consequence of having brains of finer texture, are often observed to far outstrip others in power of mind whose heads are much larger.

The brain of Cuvier, it is said, was distinguished for richness of convolutions, as well as for its weight. Mr. Grote's brain was reported, shortly after his death, as being peculiarly formed, and very rich in convolutions :—

"After his death, Mr. Grote's skull was, in accordance with his own wish, opened by Professor Marshall, and, contrary to general belief, the brain was found to be astonishingly small and of peculiar formation. It is, however, said to be very rich in convolutions. I am told that phrenologists will not be able to derive any comfort from the result of Professor Marshall's examination."—Correspondent of the *Dundee Advertiser*.

Why not, Mr. Correspondent? Did it never occur to you, or to your informant, that size is only one measure of power, and that the *peculiar* formation and richness of Mr. Grote's brain might account for his remarkable endowments?

I have examined the bust of Mr. Grote in Westminster Cathedral, and found nothing in its form opposed to phrenology, but much in support of its doctrines as a system of character.

Whether this bust be a faithful delineation of Mr. Grote or not, the artist has most effectually portrayed the mental characteristics for which he was famed, in strict accordance with phrenology. The forehead from the ear is long, and all the organs about the median line, from the top of the nose upward, are extremely large,—those organs which markedly indicate acuteness of perception; a desire for, and remarkable aptitude to acquire knowledge of the history of individuals and nations; and likewise for collating facts, for critical analysis, and classification.

CHAPTER VII.

ON THE TEMPERAMENTS.

THE mental functions are greatly modified by disproportionate organic development; for, as in the case of the brain, size is a measure of power of every organ. Preponderant size, with the consequent energy of action of particular groups of organs, and the specific effects this produces in the animal economy, constitutes the basis of the temperaments.

Hippocrates believed the higher class of animals to consist of four elements, namely,—blood, lymph, yellow bile, and black bile; and he defined the temperaments according as any of these predominated in an individual; first, there was the *Sanguine* temperament, produced by a predominance of blood; next, the *Lymphatic*, caused by an excess of lymph in the animal tissues; third, the *Choleric* temperament, resulting from a superabundance of yellow bile; and fourth, the *Atrabilious* or *Melancholic* temperament, produced by black bile. The latter element, however, had only a fancied existence, and a belief in it has been given up. Various views have been held regarding the temperaments since the time of Hippocrates, but his classification continued with little modification to prevail up to a recent period. Dr. Spurzheim adopted it, but added thereto the *Nervous* temperament.

Dr. Thomas, a French author, promulgated another theory which has gained many adherents. Believing that every organ acts with a degree of energy in proportion to its size, other conditions being about equal, he divided the organs into three groups or systems, which he viewed as producing all the vital energies of the frame. His first group, which he believed to produce the lymphatic temperament, are situated in the abdomen, and their office is to digest food and make blood. The second group comprises the heart and lungs ; they are located in the thorax or chest, and their functions are to circulate and purify the blood : they form the sang uine temperament. The third group is situated in the skull, and consists of the cerebrum, cerebellum, medulla oblongata, and cephalic ganglions, or the entire brain, which give rise to the nervous tempera ment. On the different degrees of the development of these systems, and the proportion they bear to each other, depends the temperament of any individ ual,—that is in accordance with Dr. Thomas's theory.

This classification is based on sound principles ;

but is nevertheless incomplete, as it does not include one very important class of organs. No allowance is made for the influence of the athletic structure, in which the vital influence is frequently concentrated. It is obvious that the functions which exhaust the vital force must be taken into consideration, as well as those which generate it.

It is very difficult, if not impossible, to form such a classification of the temperaments as to embrace all the disturbing influences that the body and mind are subject to, and to which objections might not be taken. They might be divided into Anatomical, Physiological, and Pathological ; and these, again, into numerous subdivisions, which, from their complexity, would have a tendency to bewilder the mind of the student, rather than enlighten his understanding.

The best method of classification, in my opinion, is the one struck out by Dr. Thomas : to group together those organs that have a similarity of function, and give names which have a direct reference to them,—indicating their power to act, rather than modes of functional activity. The following names appear to be appropriate :—First, the *nutritive* temperament ; second, the *sanguine ;* third, the *mental,* and fourth, the *muscular.* When the abdominal, thoracic, cephalic, and muscular groups are fully developed and suitably proportioned, the temperaments are equally balanced ; and individuals so constituted are most highly endowed for performing all the duties of life,—as all the parts, possessing equal vital energy, are equally fitted for exercising their functions.

The external indications of the *nutritive* temperament are a relatively large abdomen ; an inclination to corpulency ; roundness of form ; soft, flabby, inelastic muscles ; fair skin ; hair of fine texture ; light-coloured, hazy, sleepy eyes ; an indolent, inexpressive countenance ; sluggish circulation, and a general want of vivacity.

The *sanguine* temperament is indicated by a relatively large chest ; moderate plumpness and firmness of flesh ; quick, animated, expressive eyes, of blue or grey colour ; florid complexion, and vivacious countenance ; flaxen, light-brown, or sandy-coloured hair ; and a full, bounding pulse.

The *muscular* temperament is known by proportionately large muscles, of considerable hardness and elasticity, thinly covered with fat ; high cheek-bones ; harsh features ; strong, dark hair ; black or hazel eyes ; an olive-tinted skin ; and a cool, calculating manner.

The *mental* temperament is characterized by a proportionately large head ; soft, silky, dark, scanty hair ; thin, soft skin ; small muscles, unencumbered with fatty substance ; great sensitiveness ; small features ; pale face, indicating delicate health ; and bright hazel eyes, which often sparkle with vivacity and penetrating keenness.

The preceding description is a merely suggestive

outline.*   Pure temperaments are spoken of by phrenologists. Dr. Spurzheim says, " These four temperaments are seldom to be observed pure and unmixed ;" and Mr. Combe observes, " The different temperaments are rarely found pure." Certainly not. The condition is an impossibility. A combination of the temperaments in some form is a necessary condition of existence. All that can be done is to point out the indications of predominating temperaments ; but some of these vary considerably in their characteristics. Rules may be laid down to assist the student in his observations and inductions, but he must learn to observe and think for himself. The temperaments mix in all possible proportions, and are named from their particular individual combinations, always commencing with the predominating one, and naming the others in rotation, according to their size. For example, if the mental temperament be the largest, the muscular next in size, and the other two be equal, the individual would be said to have the mental-muscular temperament. My method is to register them by a scale of twenty-four numbers. The full development of each temperament is represented by the figure 6, and a higher number than 6 indicates over-development, and less than 6 deficient development. For example, take an individual whose temperaments are 8 parts nutritive, 8 sanguine, 4 muscular,

* Those who desire a fuller description should consult " Phrenology and How to Use it,"

and 4 mental, each of the two former would be over developed two twenty-fourths, and the two latter would be deficient two twenty-fourths; consequently, he would have the nutritive-sanguine temperament.

The elements of the temperaments are probably hereditarily embedded in the constitution in determinate proportions; but they are not unchangeable from birth to old age. The worst combination is capable of improvement, and the best is liable to deterioration. Though the children may groan under the punishment due to their parents' guilt, their case, happily, is not irremediable. There is a way of escape from intellectual, moral, and physical degradation; but it is conditional, and the conditions embrace self-knowledge and self-denial.

Perfect organic equilibrium is the great desideratum; and to attain this state necessitates a rigorous course of discipline, a keeping of appetite and palate under complete subjection, and an entire avoidance of everything calculated to have an adverse effect.

Individuals of the nutritive temperament should accustom themselves to physical exercise, and guard with a jealous eye the desire for indulgence of appetite, or for lying immersed in feathers, or on soft couches. Work, work, fresh air, and abstinence, should be their motto.

Those who have the sanguine temperament have a tendency to passional excitement; they should put the curb on, and constantly wear the breaking-reins. And if the inheritors of mental temperaments

wish to avoid doctors' bills, nervous dyspepsia, hypo-
chondria, monomania, nervous tremors, and the
numerous ills arising from excessive brainal activity,
they must get out of their libraries, and be more
social and jolly : they must not be afraid to laugh
and enjoy life out of the literary atmosphere. The
wants of the stomach, the lungs, and the muscles,
must be attended to with strict observance.

The practical uses of the temperaments are many
and important. In selecting individuals for particular offices, the temperaments must ever form an
element in adaptability. A man having a large
brain, but a small thorax and abdomen and thin
muscles, might possess every qualification necessary
for satisfactorily performing the duties of any office
requiring great talent and tact, extensive knowledge, or mechanical skill ; but he would not be
able to undergo great physical exercise, nor to bear
protracted bodily fatigue,—such, for example, as is
required of a general and commander-in-chief of an
army engaged in active hostilities. He might manifest consummate ability in planning a campaign, and
in strategetic manœuvre in handling his troops ; but
he would work off vitality quicker than his recupera-
tive apparatus could supply it, and consequently he
would succumb. Instead of his brain being well
supported by vigorous nutritive and circulatory
powers, it would be heavily taxed to keep these
functions active, and its stores would be thereby re-

duced, and rendered inefficient to support the energies of the mind.   Such a person, therefore, should
not be selected to command large bodies of troops
during a long and trying campaign, for he might
break down at a critical juncture, and hazard irre-
coverable defeat.   An individual who has a capacious and fully developed chest, whatever may be his
mental qualifications, is not fitted for a sedentary
occupation where he would be much confined.   His
large lungs and vigorous circulation would cause a
restless longing for a more active sphere of labour,
which, if not gratified, would probably terminate in
disgust for the office.

CHAPTER VIII.

ON THE SCIENCE OF MIND.

THE science of mind is no longer confined within
the sphere of metaphysics. Its boundaries are extended, so as to embrace the whole constitution of
man : his physical structure, each individual organ,
its connections and functions, and modes of activity.
But this Treatise has a less pretentious aim. Its
main object is a concise exposition of the science of
character : the diversities of mankind and the distinguishing characteristics which constitute mental
individuality, and the organic signs that indicate
the traits peculiar to each.

Mind is distinguished by its phenomena alone :
its essence being beyond the reach of man's perceptions it defies philosophic research, so that all
opinions on the subject are purely speculative. In
fact, the essence of matter equally eludes our grasp.

Matter and mind are opposites, and are spoken of as distinct worlds—the external world and the internal world ; but more properly as the objective and the subjective worlds.

The distinguishing property of the objective world is extension,—both as relating to resisting matter,—as a block of wood or stone, water or air ; and that which is non-resisting, or empty space. The subjective world is made up of self-consciousness and of our experiences of things, so-called, that cannot be said to be extended,—such as pleasure and pain.

The attributes of mind are divisible into three orders :—*first*, Feeling ; *second*, Volition ; *third*, Intellect.

Feeling stands first in the list of the distinctive marks of mind, and is the common property of the brute creation and of man. It includes all pleasures and pains.

States of feeling are multifarious. Warmth, coolness, an ache, relief, fatigue, freshness, indifference, joy, sorrow, love, hatred, fear ; smelling a fragrance or a stench ; being moved by the sense of beauty or of ugliness,—are all modes of feeling.

Volition : all beings possessing a mental nature can not only feel but act. To put forth an effort for the attainment of some specific end,—such as to satisfy hunger and thirst, to expend exuberant freshness, to rest, to enjoy the picturesque beauty of a landscape, to flee from danger, to bestow alms, etc., are modes of activity, and marks of mind distinct from feeling, but resulting from it.

Acts are Voluntary and Involuntary. The preceding are voluntary. Breathing and sneezing ; the process of digestion, the circulation of the fluids, the secretion of bile, etc., are involuntary operations, and are not recognised as mental, although many of them can be accelerated or retarded by volition.

Voluntary and involuntary movements are also designated as conscious and unconscious acts, and the latter are likewise called reflex actions—that is, when an act is performed for a specific aim, such, for example, as turning to relieve oneself, rubbing any part that tickles, or removing anything which gives annoyance when we are asleep. Now, when a person performs any of these, or similar acts for a purpose, the act would appear to be the product of conscious-will power, stimulated by a sense of uneasiness or pain. Yet, it often is not so ; and no record of such a feeling, or of the act consequent on it, is made in the person's mental history. What, then, are the distinctive marks of conscious feeling which results in an act of volition and those of an involuntary act that is prompted by unconscious feeling ? The former state is perceived by the mind through or by means of a molecular action of the cortical substance of the cerebral hemispheres, and the efforts put forth in the search of relief is direct-

ed by the intellect. The latter is an effect produced
on a nervous ganglion, that is not transmitted to the
cerebral hemispheres, and is thus kept outside of
the recognised sphere of mind ; and the resultant
movements are said to be instinctive operations of
the law of conservation inherent in the animal
economy.

Thought : under this head are included by the
schools the general attributes of memory, perception, conception, reason, judgment, the will, and
imagination. But all the intellectual faculties are
resolvable into two orders, viz. :—memory and discrimination.

Memory is the retentive and reproductive power
of the mind ; and necessarily stands first in the
order of intellectual functions ; for, if we did not
remember an ideal impression of our sensations, we
could not distinguish the difference between any
two or more of them by contrast, inasmuch as the
last and present one would solely occupy our attention, and no judgment of its similarity or dissimilarity with preceding ones could be formed either
in kind, intensity, or degree. A burn would not
be cognised from a genial warmth, or the tornado
from the zephyr breeze, and experience would have
no signification. The power of retaining an idea
depends on the development and the impressibility
of the sensory system, the intensity of the stimulative
force, the healthy condition of the special nerve or

nerves acted on, the strength and diffusion of the nerve-currents, and the vividness with which the idea was first perceived. These include temperament, constitutional quality, and cerebral conformation.

The diversity observable in the retentiveness of different persons is very remarkable; not only so, but persons are rare whose memory is equally retentive for all things. The memory of some seizes hold of particular facts with a firm, enduring grip, but some other facts it handles tenderly, and soon permits them to escape; and some memories resemble a sponge: though their holding capacity may be great, a little pressure of time empties them. Such memories may be further likened to a leech that voraciously feeds to satiety, and forthwith disgorges its meal. Many diversities are accounted for by phrenology; but the best observers have found something regarding the power of memory unaccountable. There is evidently a basis of retentiveness, the physical signs of which have yet to be discovered.

REASON AND JUDGMENT.—The powers by which we examine evidence, and draw inferences, and come to conclusions, imply discrimination : the perceiving of similarity and dissimilarity. The power of discrimination constitutes a feature of leading prominence in the intellectual status of man ; and an endeavour to estimate it should form one of the chief aims of the character analyst.

L

Whether we direct our attention to the outward observation of things and their qualities, numbers, positions, relations, and bearings to each other, or to their movements, or analyse the operations of our mind by introspective examination, differentiation or agreement will be found to be the main object of our research which requires discriminative capacity.

Generalization can be carried no further than this division of the powers of the intellect.

Great individual differences, however, of memory and discrimination, are everywhere observable, and they are also very numerous. To account for these differences, and the vast variety of intellectual aptitude manifested,—or to point out the outward or visible signs of this state of things in the human form,—is what phrenology aims at. This, it must be admitted, is laudable.

Gall and Spurzheim divide the intellect into two genera: Perceptive and Reflective Faculties. The perceptive faculties are said to take cognisance of the nature of our sensations : the difference between the kinds of them and of their qualities. They are likewise often spoken of as the observing and knowing faculties, and as forming the basis of the practical, matter-of-fact, and scientific order of mind when they are predominantly active.

The functions of the reflective faculties are essentially to perceive similarity and dissimilarity, and the relations of cause and effect, and to generalize their observations and deductions. When they are dominant in power, they form the basis of the theoretical, logical, and philosophic type of mind.

It is a demonstrable fact, that there are cranial signs indicative of the thoughtful and the reflective type of mind, and likewise of the mind that is more given to observe tangible qualities than abstract principles. Nevertheless, the terms perceptive and reflective do not so definitely designate these mental types as could be wished; for all the intellectual faculties both perceive and reflect. To reflect is to look inwards. The mind, as it were, reflects back upon itself and examines its own perceptions. Every faculty of the intellect is reflective so far as revolving inwards is characteristic of reflection, inasmuch as none can look outwards. They are restricted to the perceiving of sensations. To analyse, to classify, and to trace causation are the functions of the so-named reflective faculties,— namely, Comparison and Causality; which imply the perception of likeness and unlikeness, and of the relations of cause and effect; and that these mental processes are as purely perceptive as the discrimination of the difference of configuration, dimension, gravity, colour, etc. When a person endeavours to compare any of these qualities of things that are present with those that are absent, the faculties of Form, Size, Weight, and Colour reflect as certainly as Causality does when the relation of cause and effect engages the attention.

Scientific and Philosophic, or Practical and Theoretical appear to me more appropriate terms to distinguish the dissimilarity of the functions of these groups of faculties than Perceptive and Reflective. But as these names have been so long in use, and as their meanings are generally understood, the substitution of others now would probably be inconvenient.

THE SENSES.—*Touch, Taste, Smell, Hearing,* and *Sight,* are the channels through which we derive our knowledge of the objective world. Their respective nerves are adapted and susceptible to particular kinds of irritation from without, and when any of them is irritated by its appropriate stimuli, the irritation is conveyed to the brain, and there produces sensation, and when made with sufficient force, the mind becomes conscious of it; then the intellect comes into play and cognizes the nature of the sensation, and refers it to the outward cause from which the stimulus originated. The organs of sense, then, are nerves exposed to the influence of special agents.

TOUCH.—The nerve of touch is ramified over the skin and part of the tongue, by which we feel resistance or solid pressure.

TASTE.—The ramification of the nerve of taste is on the tongue. It is susceptible to certain chemi-

cal properties of liquids and of solids that are dis-
solvable in the mouth.

SMELL.—The olfactory nerve is spread over the
pituitary membrane, situated in the nose.   It is
adapted to the stimulus of gaseous effluvia, and
puts us in relation to things at a distance as well as
those that are near.

HEARING.—The auditory nerve, which is located
within the ear, is suited to the influence of aërial
pressures, or atmospheric waves.

SEEING.—The optic nerve is adapted to the rays
of the sun and other luminous bodies.   It is ramified behind the eye, and this ramification is named
the retina.

These organs of sense are named the external
senses, in consequence of being exposed to external influences, and to distinguish them from the
organs of sense that are directly effected by internal stimuli,—named the internal sense, and the
muscular sense.

THE INTERNAL SENSE.—The nerves of this sense
are ramified over the internal viscera to give due
notice of their condition,—such as hunger and
thirst, and when the excrementitious organs require
relief.   They are the media of numerous pleasures
and pains.

THE MUSCULAR SENSE.—Sensory nerves are distributed to the muscles along with the nerves of motion. Their functions are to make known to the mind the muscular condition,—such states as tension, flexion, dead-strain, fatigue, freshness, and all abnormal effects.

CHAPTER IX.

CRANIAL SIGNS OF CHARACTER.

PHRENOLOGISTS treat of the cranial signs of character as organs of primitive faculties of the mind,
and I do so likewise in "Phrenology, and How to
Use it in Analyzing Character," and have followed
a similar course in the preceding pages, especially in
the controversial sections ; but I deem it advisable,
at this stage, to depart from this method, and shall
now speak of signs of character, and define their
localities : how to estimate their size, and interpret
their indications : how to infer the talents and aptitudes for special vocations, the inclinations, temper,
and disposition of individuals : what a person may
become under certain conditions, and what he would
be more or less likely to excel in by application and
proper direction of his powers. These subjects,
however, can only be briefly treated, for the work has
already considerably outgrown the original design.

The signs are classified under two orders : the
Feelings and the *Intellect*. They are divisible again

into those common to man and animals, and those peculiar to man.

It is impossible to define the point where the animal ends and the man begins; but, roughly speaking, a line of demarcation may be prescribed in the human skull or head immediately underneath the frontal and parietal eminences. It has been observed that man is highly endowed in proportion to the mass of brain above this line; and that in persons in whom it is predominantly developed, all the moral attributes that distinguish human from brute life are dominantly active: that persons whose brains are most developed in the basal region are more strongly inclined to animal gratification than the preceding class, in the ratio of the preponderance in the size of the mass of brain below this division line to that above it.

Instructions will be given in another section how to estimate the size of the signs of character; but it may be serviceable before defining their localities and indications to note that the sizes of the signs situated in the occipital region of the head may be estimated by the extent of their backward projection beyond a vertical line drawn over the middle of the mastoid process, and those in the frontal region by their advancement before a vertical line drawn from the centre of the zygoma; and those situated in the middle region by the extent of space between those vertical lines. The width of the head gives their lateral dimensions.

CRANIAL SIGNS OF CHARACTER.

Order I.—SIGNS OF THE FEELINGS : THEIR NAMES,
NUMBERS, AND SITUATIONS.*

1. *Amativeness* is situated in the cerebellum ;
2. *Philoprogenitiveness* in the infero-posterior portion
   of the occipital area ; and
3. *Inhabitiveness* lies immediately above it ; and
   next in order above this sign is the seat of
3A. *Continuitiveness.*
4. *Adhesiveness* is located at each side of the last
   two signs.
5. *Defensiveness* is situated behind and above the
   mastoid process, on a line with the upper half of
   the ear, in the post-squamoso-sphenoid area ; and
   behind and partly underneath its lower portion is
E. *The Centre of Energy.*
6. *Destructiveness* lies about the ear. The top of
   the ear marks its superior border, and it extends
   backward to a vertical line drawn from the posterior border of the mastoid process, which separ-
   ates it from Defensiveness.
6A. *Alimentiveness* is in the squamoso-sphenoid area,
   about an inch in front of the ear. Its superior
   boundary is on a line with Destructiveness, and
   it dips into the middle fosse of the cranium.
   In front of this

---

* The student, in studying the description of the phrenological map,
should either have the model bust, or the engraving of it, before him to
refer to.

C. *Bibativeness* has its location in the same area.
  What the cerebral mass situated in the ali-sphenoid area, and that lying between it and Bibativeness indicate, is not known.   I am, however,
  inclined to think, after making hundreds of observations of the size of this part, and the health
  of various persons, that it indicates particular
  states of the abdominal viscera.   I have observed
  that persons, as a general rule, whose heads are
  very narrow at this part, are apparently predis-
  posed to sluggishness of the liver and bowels,
  and that it is *vice versa* in the case of those who
  have an opposite cranial conformation.

7. *Secretiveness* is located above Destructiveness
  and Defensiveness in the infero-parietal lobule
  area.

8. *Acquisitiveness* is directly in front of the last sign.
  It has

9. *Constructiveness* for its anterior neighbour.

10. *Self-Esteem* is at the top of the back-head, at
  each side of the median line, in the posterior
  portion of the upper parietal lobule area ; and

11. *Approbativeness* lies at the sides of it.   Then
  comes, in a lateral and anterior direction,

12. *Cautiousness ;* in the centre of which the parietal eminence is situated.

13. *Benevolence* is at the top of the brow, posteriorly
  at the sides of the median line, and in the upper
  posterior part of the supero-frontal area.   Behind it is the location of

14. *Veneration*, and next in order comes
15. *Firmness*, at each side of which is
16. *Conscientiousness.* Next it anteriorly is
17. *Hope*, at the lateral sides of Veneration.
18. *Love-of-Change* (*Marvellousness*) is in front of the lower half of Hope.
19. *Love-of-the-Picturesque* (*Ideality*) is situated laterally to the last sign, and extends a little farther back. Posterior to it is
19B. *Sublimity*, which extends to Cautiousness.
20. *Imitation* lies between Veneration and Love-of-Change, and in front of the upper half of Hope.
21. *Humorousness* is in front of, and in a downward and lateral direction to Imitation.
36, 37, and 38. *Graveness*, *Laughter*, and *Awe* are situated immediately below Ideality and Sublimity, and stand to each other from before backward, in the order as enumerated.

Order II.—Signs of the Intellect.

22. *Individuality* is situated between the eyebrows; and the following come next in order after it in a lateral course, around the orbital arch :—
23. *Form.* 24. *Size.* 25. *Weight.* 26. *Colour.*
27. *Locality* is directly above Nos. 24 and 25, which takes a slightly oblique, lateral direction from below, upwards.
28. *Number* is located behind the external angular process, and

29. *Order* adjoins Colour.
30. *Eventuality* is in the middle of the forehead, above No. 22. The upper part of No. 27 is next to it outwards ; then comes
31. *Time*, above Nos. 26 and 29, and
32. *Tune* is external to Time on the same plane. Its centre is about the line of the temporal ridge.
33. *Language.* Prominence of the eye, the amount of space between it and the supra-orbital ridge, and the fulness of the under-lid are marks of the retentiveness of the memory of words, and the talent for articulate language.
34. *Comparison* is situated immediately above No. 30 in the centre of the upper part of the brow, and
35. *Causality* adjoins it at each side, and is bounded laterally by No. 21. The frontal eminence marks its centre.

Nos. 22, 23, 24, 25, 27, 30, and 34 are in the supero-frontal cranial area ; 26, 29 and 31 in the middle ; 28 is in the inferior area ; 32 is partly in the middle and infero-middle frontal areas ; and 35 is partly in the superior and middle frontal cranial areas of the præ-coronal region. See plate No. 4.

In studying the distinctive features of the emotions, and the part each plays in the drama of life, it should be borne in mind, that difference forms the basis of individuality of character, not negation; just as dissimilarity in form and feature constitutes personal identity.

" Have I got this or that organ ?" is a common question asked by persons on consulting a phrenologist.   It would be quite as rational to ask, " Have I got a nose, or a mouth, or a chin ?" for, as these prominent features are the common inheritance of mankind, so are the whole of the emotions and talents of the human mind common endowments. As the nose does not form the face, neither does an emotion or a talent form the character; but, as the form and size of this feature individualises, to a certain extent, the face, so do a strong emotion and a vigorous talent partly characterise the mind : and as there is a vast variety of facial forms, so do heads present a great diversity in configuration; and the emotions and talents are found to be equally dissimilar.

Though we breathe through the nostrils and mouth, these organs do not indicate the measure of the breathing capacity.   Now, every person is susceptible to sensations through the nervous system, which give rise to emotions of love,—love of children, for example ; yet, this neither constitutes equality of susceptibility, nor the intensity of the emotion, nor the measure of its abiding strength and activity.   Such differences are not disputed ; but their causes, and the outward signs of them, are subjects on which disagreement exists.

Phrenologists hold that, other conditions being *about* equal, in different individuals, the comparative development of the inferior posterior portion of the occiput indicates the comparative strength of the love of children possessed ; and that all the emotions and talents have their signs in the head, from which their power can be inferred with a tolerable degree of accuracy.   Is this so?   It is needless to say the position is disputed ; and I am free to admit there is a foundation for dispute ; but it wants solidity.   I will not say it is a sandy foundation, but am prepared to prove that it will not support the structure that it is represented to bear.

Taking the love of children, as an example, by way of illustrating the case, let us endeavour briefly to expose the basis of opposition to view in the best possible light, in order to investigate it, and to form an unbiassed and correct judgment regarding its solidity.

First, then, persons having a moderate development of the sign of Philoprogenitiveness, are frequently observed tenderly nursing children, and otherwise manifesting attachment to them.   They attentively stand by the cradle, and guard it from interruption, when the health of the dear creature happens not to be up to the mark, and should it be pained, they put forth all their energies to soothe it and assuage its sufferings : they spare no time or money, or personal comfort.   Have we not, then, in such cases, every mark of strong love of children?

I decidedly answer, no. Kindness, compassion, tenderness, and self-abnegation in the discharge of duty there are markedly shown, but the prominent features of ardent love are not. What are these? Grief, nervous apprehension for the child's recovery, lest some fatal symptom may supervene, causing a fearful looking forward to the harrowing scene of separation, and an enervating, sorrowful, foreboding realization of the fact. These are but a few of the symptoms originating in predominating energy of the feeling of child-love; and should the dreaded event occur, a will-not-be-comforted sort of frenzy follows. Who can gauge the disruption that such an event causes in such a person's mind, and the endurance of the chasm; and its susceptibility to be disrupted, when the wearing apparel or playthings of the departed one turn up, or when infants that were born about the same time as the deceased are seen hanging on their mammas' bosom, or while their engaging prattle is listened to, or their winning ways beheld? The calculation baffles the intellect, and even imagination fails in the attempt to conceive it. But kind and dutiful parents calmly giving up their children at the summons of death, and reconciling themselves to their bereavement with the knowledge of having done their duty, is a scene of ordinary occurrence. These diverse manifestations are marks of dissimilarity in the vigour of the innate love of offspring. Although dissimilarity in the emotions of Defensiveness, Self-

Esteem, Cautiousness, and Hope, in different persons, modifies the effect of the feeling on the mind
of each.

It will be found that nurses who have a large development of the sign of Philoprogenitiveness experience a very different kind of enjoyment in fondling children to those whose sign is much less
developed, notwithstanding the latter may nurse
with considerable tenderness.   The former feel
thrills of pleasure succeeding each other, that come,
as it were, from every fond embrace and endearing
look.   The latter nurse not so much for the love of
nursing, or the actual pleasure they feel in it, as
from a sense of duty, and a benevolent intention to
confer some benefit on the helpless wee things, and
to make them happy.   They are actuated more by
Benevolence than Philoprogenitiveness, and would
nurse a weak, suffering adult as kindly and faithfully
as a child; while the former might barely do their
duty in an adult sick room, and even that with
grumbling.   The pleasure of nursing accruing from
the emotion of Benevolence is very unlike that
which springs from Philoprogenitiveness, and it
gives rise to characteristic traits peculiar to itself.

Every emotion and intellectual faculty may
prompt to the fondling of children, and yield
pleasure.   A beautiful child would naturally ex-
cite the emotions of Love-of-the-Picturesque, and
draw attention to it.   A child whose dress dis-
plays harmony of colour would arrest the attention

of a person whose perception of colour is acute, and would command a pleasing recognition; and finely modelled symmetrically proportioned features would attract large Form, and beget a pleasurable response.   Even a sensitive touch conveys pleasure when the velvety skin of an infant is felt.   An analysis of these various emotional and intellectual pleasures, excited by the child, shows that, while they all add to the general effect, each is merely satisfied by the quality which is its natural stimulus, and that real child-love, as a natural instinct, springs from Philoprogenitiveness specially.

Supposing two persons possessing all these other qualities in equal strength and culture, but the one having the sign of Philoprogenitiveness small, and 'the other having it large, love of children would be very differently displayed by them on such.   It is differences, however, that phrenology as a system of character rests as its basis.

Dominant emotions and talents individualise the character, as particularly formed features signalise identity.

It may be advantageous to examine another sign of the social group on which misapprehension is shown,—namely, Adhesiveness.   Strong friendship and abiding attachment are manifested by persons in whose heads this sign is predominantly large. They do not welcome you by putting their hands into yours like a dead fish, nor present a finger to

M

shake as though it were a disjointed member. They
seize your hand with a firm grip, and giving it
a lively shake, send a gush of friendly greeting
through the system.   There is no mistaking its
character : it is warm, frank, thorough and attrac-
tive.   This is their general rule of conduct ; but
exceptions occur.   Largely developed Approbativeness, Cautiousness and Secretiveness, combined with
deficient Defensiveness and Self-Esteem, produce
modesty and bashfulness, and counteract the out-
flow of friendship.   Notwithstanding, though the
presenting of the hand may not be so bold, nor the
grip so strong, still you get the whole hand in your
palm, and a discharge of true friendship into your
soul, that is nothing the less genial because modest-
ly given.   Pent up friendship is like water in a
reservoir.   You have only to prick its sides, and
out it flows by force of its own gravity.

Reader, did you ever feel the repelling shock of
a dead-fish-like shake of the hand?   What a freez-
ing chill !   How it blights the bloom of friendship.   The bare thought of it is enough to make
one shiver.

But what of the digital shake ?   It is a disjunc-
tive conjunction.   It marks the temperate zone of
friendship—a soul that is neither hot nor cold.   I
should as soon expect a shower of gold bursting
from a thunder-clap, as a large sign of Adhesiveness
in the craniums of such hand-shakers.

Adhesiveness, however, is only one member of

the social group and friendship-family of feelings. Amativeness, in full vigour and large growth, puts a force in the hand, as well as a lustre in the eye, that the loved one alone can fully reciprocate. Benevolence likewise is productive of sociality and love ; and the intellectually disposed may seek society for the gratification of their special proclivities. But enough for the present. Adhesiveness has distinctive features that are unlike those of every other emotion.

I intended to write a chapter descriptive of the influence of the combined action of the emotions and talents generally ; but the examples already given must suffice, excepting what is advanced on this head on the indications of the signs.

No. 1. *Amativeness.*—The size of the cerebellum is indicative of the strength of this feeling, and the desire of the sexes for each other's society. When it is large the feeling is found to be correspondingly strong, and ardently solicitous for gratification ; and when this portion of the brain is small, sexual love is modified in like ratio. However, this rule, like all general rules, has exceptions, and other causes must be sought than the dimension of the outward sign to account for the more ardent solicitations of this propensity: such as functional derangement and hereditary transmission.

Some time ago I was consulted by a young married gentleman whose nervous system was ener-

vated by amative abuse.   He had only a moderate
development of the cerebellum, but was evidently
worried by lascivious desire, and had reprehensively
loose views of the marriage compact.   In answer
to questions I put to him, he stated that his father
was marked for lewdness and violation of the ties
of matrimony.   This youth had probably inherited
morbid irritability of the cerebellum and genitive
organs from his dishonourable parent.*

No. 2. *Philoprogenitiveness.*—The feeling or instinct from which springs the love of young.   Probably there is no feeling of the human mind in
which greater individual differences are observed in
its manifestations than in this ; and similar diversities are observable in the size of the sign.   Where
it is large, the feeling is energetic ; and a child, or
in the absence of one, a pet, is fondled endearingly,
in that case ; but those in whom it is small show
in their demeanour towards children that their love
for them is markedly weak.   Various degrees between largeness and smallness of the sign indicate
equally modified degrees of the activity of this
feeling.   See introductory remarks to this section,
pp. 158–61.

No. 3. *Inhabitiveness.*—A person in whom this
sign is largely developed feels more reluctant to

* See model bust for localities of all the signs.

change his place of abode (other conditions being about equal) than one in whom it is smaller. But, of course, there are many modifying conditions to be considered in analyzing character : such as size and quality of brain, temperament, and the relative dimensions and influences of other signs.

No. 3A. *Continuitiveness (Concentrativeness).*— The feeling indicated by this sign seems to get satisfaction from continuous application ; and to be disagreeably affected by want of attention, change-ableness of demeanour, and rambling. It therefore inclines the mind to constancy of pursuit for its own gratification, according to its vigour.

Some persons are ever changing from one subject to another, like a bird in a bush hopping from twig to twig; while others continue in an even course in pursuit of their aim. In the former, this sign is generally moderate in size ; and it is larger in the latter. Discursive traits, versatility and fickleness, however, are not infrequently displayed by persons having a fair development of Continuitiveness. This state of mind is indicated by other signs,—such as large Love-of-Change (Marvellousness), want of self-discipline, and an equally balanced intellect, which seems to indicate that, in such a case, each faculty is equally solicitous for exercise, and when one becomes fatigued, others start into activity, and prompt to a change of topic or avocation agreeable to their tastes.

No. 4. *Adhesiveness.*—Persons having this sign large manifest considerable strength and endurance of friendship; and the opposite is shown by those who have a small endowment of it. They are fickle mortals, and not to be depended on. They make friends as the pastry-cook makes pie-crust : the richer it is, the easier it is broken. See introduction to this section, pp. 161–3.

No. 5. *Defensiveness (Combativeness).*—This sign indicates what its name implies,—power to stand on the defensive, in proportion to its size ; and when very large, the feeling tends to originate an aggressive tendency, as well as power to battle with the vicissitudes of life, and not to tamely submit to insult and wrong. Poverty of development indicates deficient courage. Phrenologists consider that this feeling is the chief source from which courage springs, and this undoubtedly is the case ; but, though it is the principal element, it is not the only one. It may prompt to defend or to attack, but the comparative vigour and activity of Cautiousness would considerably modify its action. So would Approbativeness, Self-Esteem, Firmness, and Hope. A strong emotion of fear would incline the mind to consider the consequence of the act, in which Approbativeness would take a part with a force in keeping with its vigour, by suggesting the probable loss or gain of prestige that was at stake. Self-Esteem, if it were large, would give self-reliance ;

but, if it were small, this essential quality would be deficient. Then Firmness would sustain the mind in the ratio of its vigour, and Hope would assist according to its strength. Conscientiousness would also play its part, by bringing into view the right or wrong of the matter, and the duties arising therefrom. It will therefore be observed, that in order to correctly estimate the courage of an individual. a great variety of circumstances must be duly considered.

E. *Centre of Energy.*—Many years ago my attention was drawn to the dissimilarity of the enduring qualities of different persons which I could not account for, and I was lead to infer that it has a special cerebral basis. After close observation and minute manipulation of a large number of heads in many parts of the country, passing over a lengthened period, I discovered that persons developing great power of endurance had wide heads at the base of the posterior region, directly behind and underneath the hinder portion of the sign of Defensiveness: and that deficient development of this part, and a corresponding want of enduring power, generally went hand in hand. Several subsequent years of experience and research have confirmed these observations, and I am persuaded that there is a connection between the power of bearing fatigue and severe mental and physical strain, and the cerebral part underlying the portion of the skull described,—

the outward sign of which, for want of a better term, is named the sign of the Centre of Energy.

Individuals have come under my notice whose muscles and head were uncommonly large, yet who were suffering from premature exhaustion, and I found this sign small in them. The size of the head and muscles of others was only medium, but the Centre of Energy was large ; and their power to bear hardship, I found, was remarkable.

The history of the discovery of this sign and its indications are given in " Phrenology, and How to Use it ;" although by no means so fully as it might have been. A goodly sized volume might be written on the subject, but I neither have time nor disposition to write and publish what probably few would buy, and fewer care to read. I make these remarks in consequence of a reviewer or two of " Phrenology, and How to Use it," expressing an opinion that the history of this sign is too briefly written, and too meagrely illustrated by cases, to produce conviction of its existence. My only reply is : that I do not desire any person to take anything for granted on my dictum : Nature is everywhere present. Let those who desire conviction, observe for themselves ; and, if they choose, repeat my observations and make known the result. I am wedded to nothing but truth, or what seems to me to be truth. I should throw Phrenology to the winds to-morrow, without hesitation, if convinced that it is not substantially founded on truth.

No. 6. *Destructiveness.*—The size of this sign, and the energy of the feeling, should be measured by the comparative width of the head over the ears, with its breadth in general, and especially by the depth of the orifice of the ears, below a horizontal plane drawn backward from the superciliary ridge. The position of the ears in relation to this plane varies in different persons, and in some cases to a great extent. The ears of some are fully two inches lower down than others; and as great dissimilitude is observed between their tempers. The high-set-eared are more forbearing and less vindictive than those who have deep-set ears,—notwithstanding their heads may be equally wide at this sign. The former are like the old brimstone dipped matches that required considerable heat to ignite them; the latter are like lucifers,—a little friction is apt to set them in a blaze. The former are much quieter in movement and milder in demeanour, and do not make use of such strong epithets, etc., as the latter, other conditions being *about* equal. The sarcasms and denunciations of the deep-eared class are of the rasping sort. They do not handle their opponents with kid-gloves on, nor attack principles, and what they conceive to be impediments to their success, in a mincing manner. They go to work with a vigorous aim, like a farmer with his steam-plough and grubber. They feel as much pleasure in uprooting as planting, in pulling down as building up. They attack, with a death aim, theories, dogmas, laws, customs,—everything, in fact, which seems to them behind the march of the times or against their own interests.

The feeling that this sign is indicative of puts us in relation to the destructive forces of nature, and is an essential quality in the battle of life. Under the direction of a cultured intellect, and the control of the moral powers, it prompts to the destruction of everything that is inimical to the commonweal, and urges untiringly onward the pioneers of progress and enlightenment. But, when the feeling is acutely susceptible and energetic in action, and is precipitated by selfishness, this combination instigates to the committal of the most revolting deeds.

All calm, deliberate murderers, whose heads, skulls, or casts I have examined, have had deep-set ears. This feature is especially marked in the case of slow poisoners, who with deliberate persistency took the life of their victims by degrees.

No. 6, A. and C. *Alimentiveness and Bibativeness.*—Considerable diversity is manifested by different persons with regard to relish and appetite, other conditions being *about* equal. Some appear as though they considered eating and drinking their greatest earthly enjoyment; they display a gormandising propensity, and seem to possess a digestive capacity of equal power. Others, again, are naturally abstemious, and digest the little they take with comparative difficulty. These opposite traits are indicated by the size of the parts localized as the above signs. The various intermediate dimensions that are observable of these parts indicate corresponding differences in the vigour of relish, appetite, and digestion. All other conditions considered, I have found persons having a predominance of the sanguine temperament to be the most inclined to luxurious indulgence.

No. 7. *Secretiveness.*—The trait of character indicated by this sign is an inclination to conceal. Persons in whom it is largely developed, usually display a marked tendency to prevent the inward working of their minds, their plans, aims, projects, and actions, from being observed, or otherwise made known. But the extent to which this tendency influences the conduct of any person depends on the vigour and activity of other feelings, and of the intellect. Two or more persons may have the sign of Secretiveness of equal dimension, yet may act very diversely. One may manifest praiseworthy prudence, tact, and strategy ; another may display despicable features of low cunning. Like diversities characterise those who possess this sign in a similar degree of smallness. Some seem almost void of the power to conceal, and are reprehensibly openminded and imprudent. Others, again, although frank and transparent, act in a judicious manner. Hence, it is absolutely impossible to correctly infer how a person would be likely to act under any set of circumstances without taking into consideration the general form of his head, and the effects of the combined activity of the mental powers.

Nothing savours more of quackery than pretending to describe a person's character by saying to him that a particular organ is large, and therefore you have this or that talent, and are capable of doing so and so, or you are disposed to act in a certain manner; your organ of Self-Esteem, for example, is small, and you are diffident and unreliant. I regret to say, however, that this is not an uncommon method. Another significant sign of charlatanry is the groping of heads for *bumps !*

When the feeling of Secretiveness is strong, and is properly governed, it prompts to tactic and strategetic action, and greatly aids the sly humourist and skilful dramatist. The peculiar mental traits of the fox forcibly illustrate the characteristic features of vigorous and uncontrolled Secretiveness.

No. 8. *Acquisitiveness.*—A large development of this sign marks the acquisitive person who bends his energies to make money and to own property. The feeling indicated by it also stimulates other powers to gain what they desire. Successful merchants, and persons who have amassed great wealth, will be found, as a general rule, to have the sign of Acquisitiveness large. The strength of this feeling is in the ratio of the cranial swelling at the seat of this sign.

No. 9. *Constructiveness.*—The inclination marked by this sign is to use tools in some way or other; but whether for purposes of construction or repairing, or for works of art, depends on the direction of other desires and intellectual proclivities. It finds agreeable exercise, for instance, in plotting, or in constructing a story.

No. 10. *Self-Esteem.*—The emotion of Self-Esteem imparts self-respect, self-reliance, and dignity of bearing. When it is energetic in a person of a moderate intellect, repulsive egotism is not an uncommon manifestation. Tolerable vigour of the feeling is requisite for the beginning of new undertakings, backed up by courage, application, and knowledge. Persons who possess the sign of Self-Esteem in a moderate degree of development are apt to underestimate their powers, and to quail before ordinary difficulties, or at least to hesitate much at the commencement of the struggle, and they often fail where persons of inferior ability succeed.

No. 11. *Approbativeness.*—A large development of the sign of Approbativeness is a sure indication of an ardent longing for commendation and ambition for distinction, and an inclination to take the counsel of friends, especially where Self-Esteem is moderate. It is likewise indicative of the keen susceptibility of a person to reproof: a tendency to fancy that his merits are unappreciated, and to be

jealous of rivals.   But a person in whom this sign
is small, and that of Self-Esteem large, is apt to
feel disregardful of public opinion, and consequently
to be slow in learning to benefit by just criticism.
The feeling of Approbativeness in tolerable strength
is necessary to induce emulation and a courteous
demeanour.

No. 12. *Cautiousness.*—Whatever the primitive
function of the feeling may be, and the conditions
to which it is adapted, that this sign indicates,
Cautiousness is undoubtedly a mode of its operation.
All things considered, persons who are largely endowed with it manifest more caution and susceptibility to emotions of fear than those who have a
less endowment of it; and, as a natural consequence, they are more inclined to look forward and
estimate the probable consequences of the signs of
the times, and to prepare themselves for coming
emergencies.   This feeling in great activity seems
to call into its service other emotions and talents
for its satisfaction, and thus to originate forethought
and circumspection.

No. 13. *Benevolence.*—The size of the sign of
Benevolence must be calculated according to its elevation from the boundary line between the man and
the brute, and the length and breadth of the head
should likewise be reckoned.   The same method
should be adopted in estimating the dimension of

every sign in the upper region and the energy of
the emotions indicated. The variations in the size
of this sign that are observable in different persons
distinctly indicate,—other conditions being *about*
equal,—the degree of susceptibility of each of them
to piteous emotion, and the intensity of the inclination to put forth efforts for succouring the destitute,
for assuaging the woes of their fellows, and otherwise helping them.

Sheer selfishness may prompt persons to aid
others, and in the heads of such, the sign of Benevolence may be moderately developed ; but it is
always found of ample magnitude in those who, for
the love of doing good alone, bestow charity, and
try by other means to benefit their race. Largeness
of this sign is a mark of the true philanthropist.

No. 14. *Veneration.*—Veneration is defined in the
"Etymological English Dictionary," thus : " *the act
of venerating ;* the state of being venerated ; respect,
mingled with reverence and awe." Awe is defined :
" reverential fear, dread."

Now, the emotion named Veneration does dis-
pose us to venerate, and when it is in a high state
of vigorous activity, and is uncontrolled by an enlightened understanding, or otherwise modified by
co-acting emotions, it unquestionably disposes to
reverence with awe, and tends to produce a servile
demeanour.

It is needless to say, that the production of the

latter manifestations is not the primitive function of this emotion. But to assert that to venerate in the manner usually taught is not its primitive function would probably evoke discussion, which I do not care to do. Nevertheless, such is my opinion.

If we wish to arrive at a correct conclusion regarding the proper function of any emotion or talent, we should view a man in connection with his surroundings, with a view to find out his adaptability to them. We should begin our investigations at his birth, and watch, with a critical eye, the development of his mind,—the bud as it begins to open,—as it gradually unfolds and expands into the full-blown flower. If we, on the contrary, begin with the developed mind, when it has been subjected to surrounding influences, when it has been partly moulded into shape by training, and been accustomed to act by force of habit, we will find the investigation beset with difficulties, and barren of reliable results.

Now, obedience is required of us in all the stages of life. Without it anarchy would be everywhere : revolt and revolution would be the order of things. Obedience is the chief corner-stone of the social temple. It is practical deference to the laws of God and man. Genuine obedience to God's laws is operative Veneration in its highest form ; and genuine obedience to the laws of man, enacted for the greatest possible good of the greatest possible number, is the same virtue in a lower form.

I consider, then, that the inclination to obey springs from the emotion now under consideration, and that *Obedience* is a better name for it than Veneration, inasmuch, as it more distinctly points out the natural adaptation of man to his state of existence.

All things considered, persons who have the sign of Veneration more fully developed than others, are more inclined to obey,—more deferential and reciprocally susceptible to like demeanour. Largeness of this sign is a mark of the worshipper; and a less development of it indicates a corresponding disinclination to worship.

No. 15. *Firmness.*—The feeling indicated by Firmness is expressed by its name. A firm, determined disposition; an innate prompting to surmount difficulties, to oppose aggression, and everything which the person conceives to be inimical to his interests and the welfare of those he takes under his protection, is a certain indication of a large development of this sign. A small endowment of it indicates a conscious want of power to struggle against opposing forces and a tendency to yield.

No. 16. *Conscientiousness.*—Definition: love of truth and justice. A keen sense of honour, an ever watchful eye over the inclinations, with a determination to subdue the first rising of an unrightful

N

desire, and to strangle it at its birth, shows the feeling of Conscientiousness to be vigorous ; and in all persons that display such traits, this sign will be found remarkable in size. Diminutiveness of the sign, and a selfish disregard of the rights of others, and of the truth, are usually concomitant. However, care must be exercised in analysing character, to take into consideration modifying influences—*for* one individual may have the sign of Conscientiousness absolutely less than another, and yet may act more honourably, more truthfully and justly. The indications of these traits should be sought for in the relative size of the signs of the selfish feelings in the two heads. Many indications accounting for the difference in the conduct of different persons in regard to what is just and true might be given, but space forbids ; and the student who desires to learn the art of reading character, must learn to observe and think for himself. Should he desire more aid than is to be found in this work, he will find it in the pages of " Phrenology, and How to Use it," etc. He should likewise study the works of Gall and Spurzheim, and the *Edinburgh Phrenological Journal*. This work was published in the palmy days of Combe, and consists of 20 volumes.

No. 17. *Hope.*—Ample dimension of this sign indicates a hopeful tendency of mind ; and a lesser growth weaker Hope. The relative sizes of Cautiousness, Defensiveness, Self-Esteem, and Firmness, are found to materially modify the activity of this emotion.

No. 18. *Love of Change (Marvellousness).*—The traits of character manifested by persons having this sign large, are an ardent longing for variety, new-ness, and novelty: the romantic, the weird, and plot interest. Excessive credulity is often an ac-companiment of largeness of this sign: and in-credulity is the usual concomitant of smallness of it. In such cases there is a tendency to scepticism, and it requires the convincing force of the logic of fact to beget belief. Vigour of the feeling seems to in-duce a desire to invent,—to produce that which will give it satisfaction.

No. 19. *Love-of-the-Picturesque (Ideality).*—The desire of the mind which the emotion gives birth to that is indicated by this sign, is what its name im-plies, love of picturesqueness, a susceptibility to the refining influences of art, and an inclination to adorn and beautify. The feeling exerts an influence over other emotions, and the intellectual aptitudes. and incites them to action for its own gratification, in proportion to its energy and persistent activity. It is one of the principal emotions which originates poetic embodiment of thought. When it is ener-getic, picturesqueness and quiet beauty thrill the soul with ravishing transport; but the grandeur of wildness touches it lightly.

No. 19B. *Sublimity.*—The contemplation of beauty and serenitude begets but weakly responses in some persons, whereas the sublime aspects of nature arouse all their energies into a delirious thirst for the grandeur of the scene, that only subsides by the fatigue of drinking,—they revel in tumultuous upheavings, and the upheaved,—the craggy steep, the loch and glen, and the grandeur of wild vastness. In such moods their emotions find expression in lofty strains; and, glen-like, they echo the poetry of their surroundings.

In this class, the sign of Sublimity is found to be much larger than the sign last defined. The poetry originated by the emotion of Sublimity is as unlike that which Love-of-the-Picturesque originates as the scenes are to which they appear to be adapted, and which they are more impressed with. But the energy and activity of other emotions and intellectual powers, contribute to the different compositions, and give them a colouring which, in some measure, accounts for the dissimilarity of the styles and strains of poets.

No. 20. *Imitation.*—Man is an imitative being. His words, actions, gait, demeanour, and productions are naturally copied by others, and effect society for good or ill, long after he has been gathered to his fathers. The inclination to imitate, and the capacity for faithful delineation, are everywhere observed to be very dissimilar in different persons. Some in infancy, and through the transition of every period of life to old age, manifest remarkable imitative power, while others possess it in a very moderate degree, and are equally disinclined to copy; nay, they seem to think the practice of copying styles and manners degrading, and consequently strive at originality. The heads of the former are found to be elevated and broad at the region of the sign of Imitation in proportion to the energy of the manifestation of the feeling. The heads of the latter are lower and narrower at this part than the former, and such depressing leanness is observed to be more marked as the desire for imitating is shown to be weaker.

No. 21. *Humorousness.*—This sign (organ) was named *Wit* by Dr. Gall, and Dr. Spurzheim called it *Mirthfulness*. I prefer the appellation Humorousness; still I am not satisfied with it, and wish a more appropriate name could be found. The emotion it points out, and the desires that are the offspring of this emotion, as well as the views regarding them by my predecessors, are fully treated in " Phrenology, and How to Use it in Analyzing Character." Several varieties of Wit are analysed and examples in illustration of them are likewise given in that work, which cannot be repeated here. Both Wit and Mirthfulness are undoubtedly manifested by persons having a considerable endowment of the sign under consideration; but my opinion is, that these are only modes of manifestation when acting in combination with other feelings and intellectual proclivities, and that the primitive inclination it engenders is more of a humorous sort than witty or mirthful. Wit of the highest class,—such, for example, as that of Douglas Jerrold,—and a gay, mirthful disposition, are commonly displayed by persons whose heads show a moderate, and even a small development of it. A cast of the skull of Dean Swift is in the Edinburgh Phrenological Museum, of which I possess a copy, and this sign is small in it. I am inclined to think that Humorousness,—that is, the property of the mind, so-called,—is not a simple emotion, but that it is an intellectual quality, or that, if it is emotional in function, it has also an intellectual side as well.

GRAVENESS, GAYNESS, AND AWE.

These signs are numbered in accordance with the order of their discovery; but they naturally belong to the order of the feelings, hence are treated of out of numerical rotation.

No. 36. *Graveness.*—I discovered this, and the two following signs of character, about seventeen years ago, and subsequent experience corroborates my observations and inductions regarding their distinguishing physical and mental features. Where the size of the sign of Graveness is considerable in the head, gravity of deportment and expression, and a preference for grave surroundings, characterise the individual. Touching pathos thrills him to the core : pitches his voice in a minor key, and excites within him tender emotions, which often find vent in tears. When the emotion is predominantly vigorous it tends to produce pensiveness, and often melancholy. Smallness of the sign indicates opposite characteristics. The use of this emotion appears to be adaptation to the sorrows incidental to this life, and to induce us to deport ourselves properly on solemn occasions. It is one of the poetic group, and predisposes the artist to tinge his productions with its colouring. He over whom it dominates as naturally delineates the feeling as ducks go to water, whether his forte be music or the muses, the pencil and brush, or acting. Vigorousness of the feeling is essential to persons who emulate truthful delineation of the pathetic.

No. 37. *Gayness.*—The emotion that is indicated by this sign is the very antipodes of the last. It adapts to the lively and gay. It presides at marriage feasts, and sings the song of rejoicing at births. Graveness echoes the funeral dirge, and interestedly attends the last rites and ceremonies of the departed. Gayness acts as the safety valve to the over-be-joyed mind, by venting the surplus force in laughter. Graveness relieves pent-up grief by weeping. Some persons laugh at the veriest trifles, who are neither funny nor humorous. They laugh

at their own sayings, and even giggle boisterously
at mischief. In all such I have found the sign of
Gayness predominantly developed. Some occasionally laugh and cry alternately,—almost in the same
breath. The heads of this class show the signs of
both Gayness and Graveness to be equally large.
They are alike impressible to mirth, ludicrousness,
and pathos.

I know an intelligent and kind lady of this stamp.
She says her first impulse is to laugh at a mischief;
and often when she has done so, she has been suddenly affected to tears, and felt ashamed of herself
for her improper conduct. Her aged mother, for
whom she manifests very tender regard, once had
occasion to stand upon a chair to do something,
and, in doing so, she fell down in her presence.
Now, instead of going to assist her mother, she
was instantly seized by an uncontrollable fit of
laughter, at the ludicrousness of the old lady's position, which, having subsided, sorrow and weeping
followed suit.

No. 38. *Awe.*—That there are great individual
diversities in impressibleness by awe-inspiring scenes
and topics, is a fact of general experience ; and I
am convinced that the comparative size of this sign,
and the various degrees of susceptibility observed
in different persons, have some connection. This
conclusion has not been hastily arrived at. My researches have been diligently continued for seven-

teen years, and during this time I have manipulated a very large number of heads and skulls, and have inquired into the history of their possessors, with the aim of testing the correctness of my observations regarding this sign, and what it indicates; and I have invariably found,—other conditions being *about* equal,—a concomitance between largeness of it and acute sensibility to awe-commanding phenomena.   Persons so constituted manifest an unusual inclination to behold the awful grandeur of nature, and they describe it more graphically and poetically than others in whom the sign is smaller. They embody their emotions in terms expressive of profound Awe.   This emotion, doubtless, contributes to the production of poetry and art.   When it is very vigorous, it seems to produce fear.

Order II.—Signs of the Intellect :—their Indications.

Remarkable individual varieties of talent and intellectual aptitude are manifested throughout the world.   Equal capacity for learning, and the desire for studying every branch of scholastic knowledge, are as rarely found in a family as perfect likeness of feature; and there is as great diversity in the retentiveness of the memories of persons.   Parents, teachers and all classes, are agreed on this point.   Some persons from infancy display an almost insatiable desire for knowledge, and to make books

their constant companions. Others delight in the
companionship of their fellows, and manifest a sort
of gregarious instinct : such persons care comparatively little for book-learning. Some are diligent
observers of Nature, and study her teachings with
extraordinary ardour : some display a very opposite
taste, and if they learn anything of her works, they
get it at second hand—that is, from written or
oral descriptions. Some are thoughtful and much
given to reflection : some seldom or never think on
any subject excepting for the gratification of appetite. Then we have every variety of talent manifested for one or more departments of study in
diverse degrees of strength. One shows a speciality
for arithmetic, another for history, a third for geography, a fourth for language, a fifth for mathema-
tics, a sixth for logic, while a seventh class seems to
have no speciality : their mind ranges through the
whole field of knowledge ; and some of them succeed
in distinguishing themselves for versatility, but pro-
foundity in any branch is rarely attained by any of
them. Others strike out a course for themselves,
and pursue the object of their search assiduously.
They, as it were, run in a groove and acquire
distinction as authorities in their several departments. Some delight in the sciences, while others
are more inclined for philosophic studies, and some
display considerable aptitude for observing special
qualities of things, but possess the ability for forming correct judgments of other qualities in a less

degree.   Some have a predilection for ethics, some for æsthetics, and others for political economy, etc.

Now, phrenologists, while leaving a large margin for the effects of education, teach that prominent connate varieties of talent, intellectual aptitude, and proclivity, are indicated by corresponding varieties of cranial form ; and they recognise fourteen signs, or organs, as they are usually called, of the intellect : each pointing out a special adaptation.   The method, however, of dividing the intellectual powers into distinct faculties, and treating of them more as independent factors of the mental economy than as members of a family, each possessing distinguishing features, is objectionable, and leads to error.   The truth is, we know little or nothing of faculties.   But we do know, or may get to know, a deal about outward cranial signs and mental manifestations—feelings, memory, talents and aptitudes.   We cannot with certainty localise a mental faculty, but we can, by observation and induction, localise a sign and learn its indications ; and can learn, by attentive application, to infer character from the sum total.

No. 22. *Individuality.*—Gall observed that persons having this sign and the one now called Eventuality large, were close observers of things, their movements and phenomena, and from the rapid progress they usually made in such studies, he was led to name the part embraced by these signs, the organ of Educability.   Spurzheim's observations led him to conclude that the lower portion of the sign (organ) of Educability (according to Gall) indicated "the faculty which recognises the existence of individual beings"—that is, as whole things, without paying any attention to their qualities, and that the upper portion of it, and the talent for events, have a connection; and he gave those names to these signs that they are now known by.

That the sign called by Gall, Educability, either indicates two distinct talents, or modifications of one, is a fact capable of demonstration. Still, Spurzheim, in my opinion, failed to comprehend the nature of the talent that the so-named sign of Individuality indicates.

A thing may be cognised as a whole without knowing its parts,—such, for example, as a watch; but the idea of the mind being capable of distinguishing a concrete existence, without taking cognizance of its external qualities, is simply absurd. No wonder, then, that Spurzheim's analysis of the functions of Individuality has given rise to opposition and banter. However, criticism must end here, for reasons already repeated a few times in the preceding pages—want of space.

Ample dimensions of the signs of Individuality and Eventuality, and love of history, are concomitants. The former indicates a talent, or aptitude, for observing things minutely: for studying geology, practical chemistry, and the sciences generally in a greater or lesser degree, accompanied with a desire to understand the nature and uses of things : and in the study of grammar, the faculty of which it is the sign attends more especially to nouns. The latter points out a special aptitude for the study of events, things in action and movements in general, and all that is expressed by verbs.

Persons who possess a greater development of the former sign than the latter, are, as a general rule, fonder of reading biography than general history; but it is *vice versa* with those having an opposite development of these signs ; and there is manifested by persons so differently constituted as great a dissimilarity in their memory for these things.

Capacity for minute observation, a tendency to personification, and a large sign of Individuality, usually go together. But smallness of it is indicative of a corresponding weakness of the talent.

No. 23. *Form.*—Sign of the talent for the discrimination of configuration, and the perception of symmetry. Disproportion is disagreeable to persons in whom this sign is large, and an ample development of it is a necessary qualification for the designer and the draughtsman, and for artists in general.

No. 24. *Size.*—Sign of the talent that discriminates dimension. Like that of Form, it is requisite in tolerable energy for those engaged in construction and the arts, and for forming a correct judgment of the size of things,—for estimating of space and distance.

No. 25. *Weight.*—Sign of the talent for judging of gravity, and the power required to overcome resistance. All persons whose occupation requires manipulatory skill, special engineering, and mechanical aptitude, should have this sign well-developed.

No. 26. *Colour.*—Sign of the talent for discerning the difference of colours, and the aptitude for mixing them harmoniously. An energetic talent of colour, and the preceding ones, adapt a person for painting, and, when acting in combination with the emotion of Love-of-the-Picturesque, or of Sublimity, or both, and with Imitation, fit him for excelling in this art. A maker of artificial flowers, and a milliner and dressmaker, should possess these talents, as well as Love-of-the-Picturesque in tolerable strength and activity.

No. 27. *Locality.*—Sign of the talent for judging of the relative positions of things, and for the remembering of places. It indicates the geographical, topographical, and astronomical aptitudes ; and an inclination for travelling. Navigators, and civil and military engineers, especially require it fully developed. Pupils who are weakly endowed with the talent indicated by this sign make only moderate progress in the study of geography as compared to those who have it in larger growth. It, with Form and Size, are indicative of the talent for mathematical studies.

28. *Number.*—Sign of the arithmetical talent. Famous mental calculators have invariably had the sign of Number large. A person may have considerable capacity for calculating by the aid of graphic signs who possesses only a moderate development of the outward sign ; but the distinguishing feature of an uncommon development is extraordinary natural ability 'for solving arithmetical problems mentally. Ample size of it is necessary for bankers, accountants, and mercantile clerks. Persons in whom the talent is vigorous feel a pleasure in keeping accounts. The feeling of genuine pleasure in a vocation results from natural fitness for it, other conditions being *about* equal, as naturally as effect follows cause. Distaste for an occupation is not always indicative of a want of intellectual fitness, but often it will be found to be so. Active power of calculation materially assists the mathematician, but considerable excellence in this science may be attained by persons who possess the arithmetical talent only in a moderate degree.

29. *Order.*—Definition : the sign of the perception of order, and the inclination to sort things up, and put them aside in a manner agreeable to indi-

vidual taste. Fastidious taste for order is usually shown in the habit and demeanour of persons in whom this sign is large, and disorder affects them disagreeably.

A few years ago, a boy, about thirteen years of age, belonging to Chester-le-Street, was brought to me by his mother to get an analysis of his character. I observed that this sign was of considerable magnitude, and after making some remarks respecting it, I elicited from his mother the following account regarding his habits. That he had a place for everything he possessed, and regularly put each in its place ; and he seemed to be actuated by a painful sense of jealousy that some person would disarrange them. He often suddenly left his companions and play, and rushed into the house to overhaul his drawers, etc., and he would not go to bed at nights until he had sided his things away, and satisfied himself that all were in their proper places. One night, while he was thus engaged, his father ordered him off to bed, but he refused to go until he finished what he was doing. His father determined to compel obedience, and ultimately forced the fulfilment of his commands by corporal punishment. The boy, however, though forced to bed, was not compelled to sleep, and so he lay awake until four o'clock in the morning, when his father went to work ; and immediately he heard the door closed after him, he leaped out of bed, and completed what he had been prevented from doing the pre-

vious night.   I got his portrait taken, and regret that
I cannot give it in illustration of the form of his head.

All persons who have this sign large are not invariably neat and methodical.   Some merely keep
things clean and in their places, but manifest bad
taste in arranging them.   Some display good taste :
they fold neatly and arrange picturesquely.   This
class will be found to have the signs of Form, Size,
and Locality fully developed, as well as Order.   It
is not uncommon to find that persons, well-endowed
with the signs of Form and Size, but possessing only
a moderate endowment of Order, display greater
niceness in physical arrangement than others who
have a much larger sign of Order.

The talent indicated by the sign of Order probably takes cognizance of the order of sequence,
and exerts as great an influence in the arrangement
of ideas as it does in that of things, and thus it
contributes to coherence of thought.

No. 30. *Eventuality.*—The traits of character indicated by this sign are described under the head
of Individuality.

No. 31. *Time.*—The talent that discerns the dura-
tion of intervals, adapts us to the onward march
of time, and notes the length of periods that inter-
vene between successive events and sounds.

No. 32. *Tune* or *Sound.*—Sign of the talent which

o

discriminates the difference of sound and of pitch. Largeness in the development of this sign and that of Time are indicative of a good musical ear ; but alone they neither fit a person for composing music nor for excelling in any kind of musical performance. The composer requires a generally well developed intellect, and the signs of Sublimity and Love of the Picturesque amply grown.   To sing well a good voice is necessary, and adaptation for skilful mani- pulation requires the sign of Weight large.   And moreover, ability to give good expression cannot be acquired without feeling.   When a person is per- forming a piece of music, he must be moved with similar feeling, and in the same degree of intensity that the author of it was when he composed it, in order to give just expression to it ; and to arouse the like feelings in listeners.   Some musical per- formances are merely mechanical exhibitions,—the natural result of deficient feeling.   Some, again, cause successive waves of emotion to thrill the soul as a hurricane moves the watery deep.   This arises from strong feeling of the right sort acting in concert with the intellectual aptitudes.   Study and practice accomplish much, but, without natural adaptation, no artist is likely to rise to the top of his profession by his own merits.

No. 33. *Articulate Language.*—Sign of the talent for giving oral expression to feeling and thought : for committing words to memory, and command in

the use of them.   The difference in the power of
speech displayed by individuals is very marked.
Some naturally talk with an easy copious flow when
the mind is serene ; while others, in a state of calm,
speak with an effort.   Largeness and smallness of
the sign are the usual concomitants of these diverse
states.   Fluency of utterance under strong feeling
by persons having only a small sign of Language is
often observed ; and this gives rise to objections
against the localization of it.   But such articulate
manifestations are not indicative of great power of
speech, but are more the result of strong emotional
pressure.   As a small orifice in a heavily pressed
boiler will discharge more steam, and make a greater
noise, than a larger one would do in a boiler lightly
pressed, so it is with verbal utterance.   Emotional
excitement, and the individual's susceptibility to it,
should be duly considered in estimating his ability
to speak fluently in general, and on special occa
sions in particular.

A good verbal memory, ease, and copiousness of
expression, in a state of mental calm, are marks of
a good talent of articulate Language,—not rapidity
of utterance when urgently moved by feeling.

Were I treating of faculties, organs, and func
tions, this section would demand lengthened con
sideration, but my present method does not require
this, and I conclude by simply remarking that the
sign now being treated of does not point out the
talent,—or, more properly, the talents,—for the

science of Language, or Philology. A person may have an aptitude for excelling in the study of this science, and yet be a poor public speaker, whilst another possessing admirable speaking ability may be a very inapt student of the science of Language.

No. 34. *Comparison.*—This sign indicates an aptitude for analysis, for criticism, and for generalization. When well developed, and in excess of Causality, in any person, he generally reasons analogically, and seems to get information, and impart instruction, best by example and illustrative similitudes. Where the lower portion of this sign is larger than the upper, the individual is more apt in perceiving similarity than difference ; whereas those in whom the upper portion is the largest, manifest an opposite aptitude. Their eye seems to take a critical sweep, and to fix on the points of difference, and they show more discriminative judgment and logical acumen than the others do.

Comparison is not a good name for the mental trait which this sign marks, for it is calculated to mislead. Each member of the intellectual family compares the qualities it is adapted to perceive, and discriminates the difference between two or more of them. Such functions cannot be performed by the faculty now in question. The perceptive powers particularise. Comparison generalises ; hence, sign of Generalisation would be a more appropriate designation.

No. 35. *Causality.* — Sign of the talent or faculty that perceives causation, and inclines us to adapt ourselves to its operations, and to endeavour to comprehend first principles. When possessed in predominant energy, it inclines to philosophic and abstract studies. An ample development of it is essential to the logician.

Comprehensiveness and profundity need not be anticipated of persons whose sign of Causality is small, although they may be smart, intelligent, and popular.

Notwithstanding that the special feature indicated by a large sign of Causality is an inclination to trace causation, and to study first principles, and though when it acts in combination with good observing powers, comprehensiveness, depth of understanding, and logical aptitude are generally manifested by minds so endowed. we cannot rationally infer these qualities of every person who possesses this sign in excessive development. Dulness of comprehension is not uncommonly displayed by such persons. They often show a remarkable obtuseness of perception, and look at a person when he is narrating a story with a vacant stare, that unmistakeably says, "I do not understand what you say ; be kind enough to repeat it." Nothwithstanding, thoughtfulness is characteristic of them. They are given more to inward thought than to outward observation, thus showing deficient perceptive talent and preponderant reflective talent; and persons oppo-

sitely organised outrival them in clearness of perception and in the command of data, and consequently surpass them in argumentation.

CHARACTER-MOULDING. — Vocation-adaptability
and constitutional fitness ; an aim and a claim,—a
notable fact.

INFLUENCE OF TRAINING.—Education, companionship, and force of habit exert a powerful influence in moulding character, and some persons
are more especially affected in this way.   To the
case of such, the adage, " evil communications cor-
rupt good manners," forcibly applies ; and, if by
any means the peculiarity of their mental constitution could be ascertained, so as to forewarn them
of their idiosyncrasy, and likewise to direct the attention of their parents, guardians, and teachers to
it, in order to enable them to understand it, and the
means of controlling it, this would be a great boon.
For then many might be saved by judicious training,
who now make shipwreck of character on the shoals
and quicksands of improper influences.   But this is
not all.   Their powers might be so nursed and directed for the good of themselves and their kind, as to
fit them for urging onward social, moral, and intellectual progress.   This is one of the aims of enlightened and well-meaning phrenologists.   They believe
that the system of phrenology, rightly understood
and properly used, would be the means, to a considerable extent, of working out this desirable end.

The preacher frequently dilates on the frailties and besetments of the mind. The phrenologist does so likewise in another form. He views the case in its scientific aspects, and speaks of natural inclinations and the predisposition of persons to act in one way in preference to any other; and, further, he avers that predisposition is indicated by the configuration of the head, so that there is no need to wait and learn the natural proclivities of any person as they develop, and thus acquire a strength by habitual exercise that can only be overcome by almost superhuman effort. The phrenologist teaches, in this respect, all that the preacher does; and, more, he does not wait for the development of uncomely traits, but begins at once,—goes to the root of the matter, and applies himself to the developing of the small cerebral parts, and to arrest the growth of such as appear to be naturally in excess, and, by this means, he tries to bring about a healthy, moral, and intellectual balance of mental powers.

CONSTITUTIONAL FITNESS.—Again, since success in life depends on a person's own merits, it is of the first importance that his vocation and natural aptitudes should correspond. Possibly dame fortune favours some, and acts contrariwise in the case of others; but industry, fitness, opportunity, and push are the principal factors in successful achievement. That many persons fail to succeed through want of

adaptability, there need be no doubt.   Hence, it is
the duty and the interest of parents to ascertain the
natural powers of their children, and, having done
so, to put them to the vocations that they are fitted
by nature to succeed in, as far as lies in their power.

It is very disappointing to parents to find that a
member of their family is unable to make headway
in the situation in which he or she has been placed,
and to discover, after painful experience, that this
arises from want of fitness for it ; but it is more
especially discouraging to the individual concerned.

The necessity for natural fitness equally applies
in the forming of business partnerships and matri-
monial engagements.

How often do we hear of unhappy alliances and
dissolution of partnerships that have had their bases
in constitutional unsuitability.   The prevention of
such painful circumstances as the preceding is a
part of the phrenologist's mission.

In education, too, the teacher should adapt himself to his pupils.   There should be no hard and
fast line of routine applied to all, but rules should
be suited to the mind of each individual, for
every person differs in some respect from every
other person.   This, also, forms a part of the
phrenologist's aim.

Surely, then, there are not many persons prepared to call his aim in question, however much
they may be led to doubt the means he employs
to hit it, and the probability of his ultimate success.

What I ask of all, is, simply to give the principles laid down in this work an impartial, thoughtful, and thorough investigation ; and, should they be found in the least at variance with truth, to do me a favour by pointing the error out ; and I faithfully promise to give it, and the results of all right-minded investigators, a candid and unbiased examination, with the sole view of arriving at the truth.

At the present time phrenology has no legitimate claim to be a complete system of mental science. It only forms a part.   The science of mind takes in the whole man, and embodies anatomy, physiology, phrenology, and psychology ; and to get a knowledge of man, he must be studied in his abnormal as well as in his normal condition ; hence, pathology must be included.   The science of mind, then, is the science of man ; and the science of man comprehends the forces of his surroundings,—in fact, of the universe,—a very comprehensive study.

Gall and his disciples have possibly fallen into the common error of partisan propagandists, and hoisted their colours too high, and so they may have to lower them a little ; but the time for pooh-poohing phrenology is past, and that for investigation is present.

Phrenology is not dead, nor does it give any signs of decay, but manifests a vigorousness and a hardihood that portend a long life.   Its nomenclature has permeated the languages of all civilized nations, and its principles are more believed in and practised now than they ever were at any former date.

APPENDIX.

THE first part of a work on "Human Anatomy," by Professor Turner, was published a few weeks ago, in which he shows a considerable leaning towards phrenology. He makes the following remarks on the probable functions of the cerebrum :—

"As regards internal structure, evidence has already been given that all the convolutions are not constructed on precisely the same plan, and it has also been pointed out that the convolutions are not all connected in the same way with the great cerebral ganglia. These structural modifications unquestionably point to functional differences in the several parts in which they are found. But, further, special connections through the arcuate fibres are established between certain convolutions and not between others, and, it is possible, not only that particular combinations of convolutions, through an interchange of internuncial fibres may condition a particular state of intellectual activity, but that these combinations associate various convolutions together in the performance of a given intellectual act, just as in the muscular system several muscles are, as a rule, associated together for the performance of a given movement."—P. 294.

The *Scotsman*, in a review of this work, March 26th, 1875, after noting the author's bias towards the doctrines of phrenology, says :—

" A fact, however, which seems to deter the Professor from fully admitting that mental power is in strict harmony with cerebral development, is the occasional occurrence of large and heavy brains without the manifestation of any great intellectual attainments. He grants that the brains of distinguished men—such as Cuvier, Dr. Abercrombie, Professor Goodsir, etc.—showed a high average weight ; but, apparently, he looks upon this fact as neutralised by equally heavy brains having been found in the skulls of fools or lunatics. But the very rudiments of phrenology teach us that high mental attainments are not associated with mere mass or weight of brain, but with the predominance of certain of its parts over others, as well as with the quality of its tissue, and the cultivation or training to which it has been subjected. Daily experience shows us that many small men are far superior in endurance and vigour to large or bulky men ; yet no one ventures on this account to doubt that a large muscle, *cæteris paribus*, is more powerful than a small one. And surely it will not be denied that it is not given to every man to attain, under adverse circumstances, the high intellectual development which he would have achieved if he had been more favourably placed. We shall never know how many Miltons, or Shakespeares, or Wellingtons, the world has lost from the possessors of brains of excellent original quality and abundant quantity having been born into the world in circumstances that gave them no chance to acquire distinction.   Under a different training the hewer of wood and the drawer of water might have become the distinguished lawyer or professor."

Well done, Sir !   You have evidently passed the deterring stage of doubt.   If you have not yet reached the goal of belief in Gall's doctrines, you know the way, and are pushing on towards it.

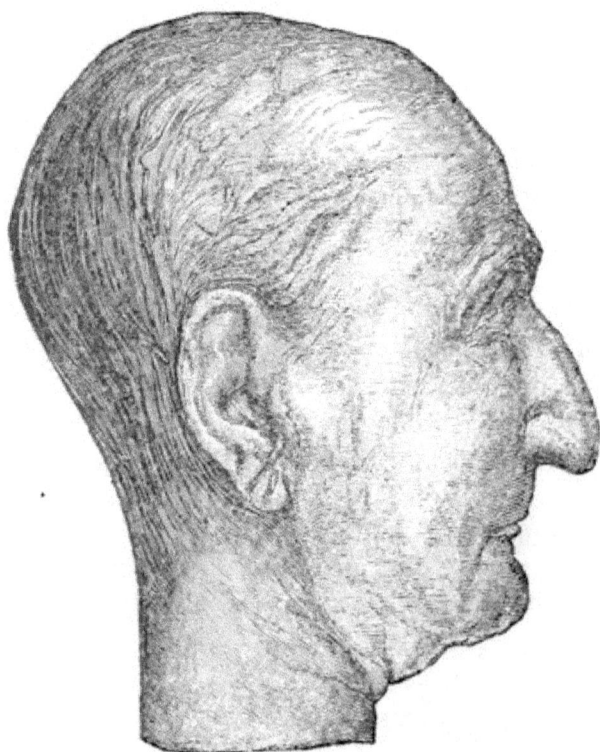

No. 15.—Rev. ROWLAND HILL.

The subjoined engravings forcibly illustrate that "high mental attainments are not associated with mere mass or weight of brain, but with the predominance of certain of its parts over others," although they were got up for a different purpose. They are drawn to a scale of one-fourth, from casts taken from nature that are in Spurzheim's collection, and which have been recently added to the Phrenological Museum, Edinburgh.

This is a massive and well-formed head—the moral region predominating. The signs of Inhabitiveness, Continuitiveness, Self-Esteem, Firmness, and Veneration are very large, and Benevolence is large.

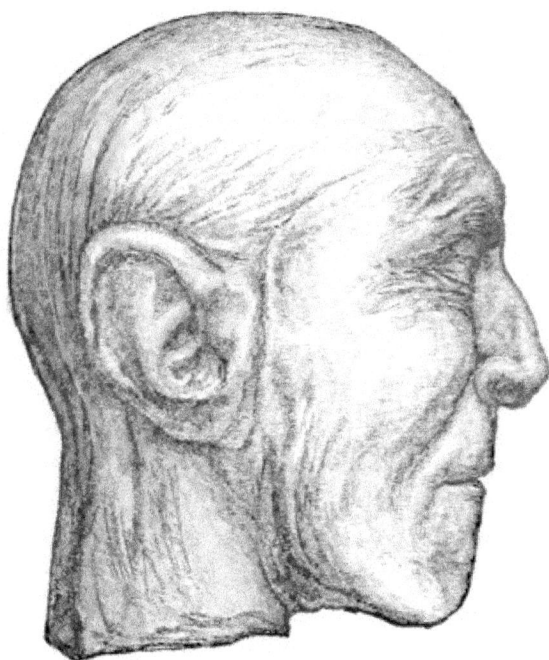

No. 16.—JEREMY BENTHAM.

Bentham's head is also large, yet it is less than Hill's; but his intellect is considerably larger, while the moral and social regions are less, especially the latter. The deficiency in Bentham's moral region is in what are called the Selfish Sentiments, or those that are common to man and animals. His sign of Benevolence is quite as large as Hill's, and that of Veneration is nearly so ; but the signs of Firmness and Self-Esteem are considerably less. This difference of cranial form indicates that Bentham would be naturally more humble-minded than Hill,—that he would incline to estimate his social status in accordance with the intellectual standard ; while the intellect of the latter would be influenced in the estimate of himself by the self-regarding principle. The intellect of Bentham is extraordinary ; the only deficiency being in the sign of Causality ; but the signs of the Perceptive Powers and Comparison,—those which indicate discriminative and analytical talent, are very massive, and considerably in excess of Hill's. And, notwithstanding that the head of Hill is much larger than Bentham's, and indicates a very capacious mind, Bentham's head is indicative of far greater intellectual capacity, depth of penetration, and knife-like acuteness of discrimination.

It is not an uncommon occurrence to find some of the compositions of authors obscure who have the sign of Comparison and all the perceptive signs large, as Bentham had. This may be accounted for in two ways : *first*, by a deficient development of Causality, and consequent incapacity of the writer to clearly perceive a subject in all its bearings that involves abstruse relations of cause and effect ; and, *second*, by his having extraordinary faculties of perception and analogical discrimination, so that that which is quite transparent to him may appear obscure to a certain class of readers. This does not arise from want of ability in the author to write lucidly, but from his not taking into consideration that all his readers are not similarly endowed, and as well versed in the subject of which he treats as himself. He presumes that what is clear to his own mind will be equally clear to them. A person having only a moderate development of Self-Esteem, but a gigantic intellect, is likely to err in this way, especially when his perceptions are very active

when he is treating on a subject involving exquisite nicety of discrimination.   Whether or not some of Bentham's writings appear cloudy, I am unable to say; but I have been told that such is the case.

www.ingramcontent.com/pod-product-compliance
Lightning Source LLC
Chambersburg PA
CBHW030734280326
41926CB00086B/1359